Making Mathematics Meaningful

For Students in the Intermediate Grades

Werner W. Liedtke

Order this book online at www.trafford.com
or email orders@trafford.com

Most Trafford titles are also available at major online book retailers.

Cover Design and Designed by Karen H. Henderson

Printed in the United States of America.

ISBN: 978-1-4269-3880-1 (sc)
ISBN: 978-1-4269-3881-8 (e)

Our mission is to efficiently provide the world's finest, most comprehensive book publishing service, enabling every author to experience success. To find out how to publish your book, your way, and have it available worldwide, visit us online at www.trafford.com

Trafford rev. 09/28/2010

www.trafford.com

North America & international
toll-free: 1 888 232 4444 (USA & Canada)
phone: 250 383 6864 • fax: 812 355 4082

Table of Contents

To Samantha, Dylan, Eliot and Lucas

Acknowledgements

Over the years many outstanding teachers were observed in action in their classrooms. Some of these teachers made it possible for selected lessons to be videotaped. Quite a few teachers became partners in a variety of action research projects. Teachers made their students available for thorough diagnostic as well as achievement interviews.

A sincere thank you goes out to all of these teachers and to the many students who were always willing to share ideas and talk about mathematics.

A sincere thank you goes to Judith Sales who read the manuscript and made valuable suggestions.

A special thank you goes to Dorothy because of her patience, encouragement, comments and meaningful questions about teaching and learning mathematics.

Chapter 1 – Fostering the Development of Numeracy

Selected Challenges

Numeracy is much more than performing operations with numerals. *Numeracy can be defined as the combination of mathematical knowledge, problem solving and communication skills required by all persons to function successfully within our technological world.*[1] Numerate persons can make sense of mathematical ideas that are part of everyday experiences and they possess characteristics that are favourable for lifelong learning about mathematics.

Fostering the development of numeracy or mathematical literacy is the major general goal of the new mathematics curriculum.[2] Implementing this curriculum presents challenges that are related to:

- Areas of sense making.
- Aspects of cognition.
- Characteristics in students favourable to the learning of mathematics.

Translating the *critical components* of the mathematics curriculum into action requires special teaching strategies and instructional settings. Assessment strategies are required that reflect the learning outcomes related to major goals of the curriculum. Reports to parents need to include information about indicators of sense making, aspects of cognition as well as characteristics favourable to the learning of mathematics.

A variety of reasons exists for the fact that students arriving in the intermediate grades will be at different stages as far as having reached the learning outcomes from the primary grades is concerned. Diagnostic settings and strategies will be required for some of these students before effective **I**ndividual **E**ducational **P**lans – **IEP**s – can be developed and presented.

Since the critical components and the goals of the new curriculum are quite different from the mathematics teaching and learning experiences that teachers and parents have had, new information needs to be shared, analyzed and assimilated. In order for parents to be able to reinforce and supplement what students experience in the mathematics classroom, important information needs to be made available to them.

The Purpose and the Parts of the Book

The main purpose of this book is to identify the important components and learning outcomes of the curriculum for the intermediate grades. Suggestions are made for teaching strategies and assessment techniques deemed appropriate for reaching the key goals and major learning outcomes. The suggestions in the book include ideas for:

- Teaching strategies, classroom settings and sample questioning techniques.
- Types of activities, problems and appropriate practice.
- Sample assessment tasks and suggestions for reporting to parents.
- Diagnostic tasks, settings and strategies.
- Reflection or discussion.

Whenever appropriate, discussions will begin with the identification of important pre-requisite ideas, procedures and skills. Examples from these discussions are suggestive of introductory teaching-learning settings and they can also be used for diagnostic purposes. The responses that are elicited during diagnostic interviews can be used for planning effective intervention – **I**ndividual **E**ducational **P**lans (**IEP**s).

The intent of the questions that are included at the end of some sections and at the end of each chapter is to provide ideas for reflection or for initiating discussions.

Theoretical comments will be kept to a minimum. Brief reference will be made to general guidelines that research provides with respect to teaching, learning and assessment.[3] Suggestions will be made for accommodating selected types of responses from students.

The suggestions that are made and the examples that are cited in the book are of a practical nature since they are based on:

- Experiences with students in classrooms.
- Observations in classrooms of teachers interacting with students.
- Action research conducted with teachers and students in classrooms.
- Responses collected from students during diagnostic interviews.
- Transcripts of diagnostic interviews with students and conversations with adults.

Challenges Related to Curriculum Content

Aspects of Sense Making and Problem Solving

One of the key goals includes the maintaining and fostering of all aspects of *sense making* in mathematics. These aspects include:

- *Number sense.*
- *Spatial sense.*
- *Measurement sense*
- *Statistical sense.*
- *Sense of relationships.*
- Developing and applying new mathematical knowledge *through problem solving.*

Problem solving is one of the major goals of teaching students about mathematics. The curriculum includes the statement that, '*learning through problem solving should be the focus of mathematics at all grade levels*' [2] (p.8). It is in these *through* or *via* problem solving settings that students develop their own problem solving strategies. New problem solving strategies may be encountered when students are given opportunities to discuss and compare strategies. An awareness of new strategies may lead to attempting these in the future.

The key to creating a *through problem solving* setting lies in how tasks are presented and how questions are phrased and posed. The onus is put on the students who are requested to use what they know and try to come up with solutions or suggestions for a solution. The following are examples of types of questions and requests that invite students to invent which is one component of a balanced and effective mathematics program:

How would you or could you ...?

Try to think of at least two ways to...

The ability to respond to these types of requests is dependent upon one or more pre- or co-requisites. These can include: *number sense*; *spatial sense*; ability to *think flexibly*; ability to *visualize*; ability to *generalize*; ability to *connect*; ability to *estimate*; high *self-esteem*; *willingness to take risks*. The accommodation and development of these aspects of cognition and these favourable characteristics has to be a component of ongoing instructional settings in the mathematics classroom.

A request that includes, '*Try to think of at least two different ways to ...*' not only serves to accommodate individual differences, but illustrates the important idea that it is better to solve a problem in many different ways than to solve many problems in the same way. This idea and more is beautifully illustrated by Calandra's story[(4)] who was asked to be a referee on the grading of a student's response to an examination question on a physics exam.

The student had answered a question about determining the height of a tall building with the aid of a barometer with the suggestion to take the barometer to the top of the building, attaching a long rope to it, lowering the barometer to the street, bringing up the rope and then measuring the length of the rope to determine the height of the building. The student appealed his grade of zero, since according to him he had answered the question completely and correctly and he claimed that he would get a perfect mark if the system was not set up against the student.

The appeal resulted in giving the student six minutes for another opportunity to record a response with the proviso that his answer should reveal some knowledge of physics. Since after five minutes he had not written anything, he was asked if he wished to give up. The response of '*no*' was accompanied by the comment that he had many answers to the problem and that he was just trying to think of the best one. In the next minute he recorded a response that made reference to dropping the barometer, timing its fall with a stopwatch and using a formula to calculate the height of the building. Almost full credit was awarded.

When the student was asked about the other responses he had in mind, he referred to measuring shadows and using proportions; swinging the barometer as a pendulum at the top and the bottom and calculating different values of 'g'; and using the barometer as a unit of measurement to count the number of 'barometer units' it takes to describe the height of the building. His final suggestion consisted of offering the barometer to the superintendent of the building in exchange for being told the height of the building. The student did admit that he knew the conventional answer, but that he was tired of instructors trying to teach him how to think in a pedantic way.

Teaching *through* or *via* problem solving requires asking appropriate questions, orchestrating follow-up discussions and accommodating different types of responses. Students are given the opportunity to develop their own problem solving strategies, to think about their thinking, and to become flexible in their thinking. Listening to others and discovering new ways of approaching or solving problems can result in advancing students' thinking. The success of orchestrating the complex instructional setting that is required for reaching these desirable results is dependent upon the important role played by a skillful teacher.

Aspects of Cognition

The fostering of aspects of *cognition* related to learning about mathematics includes:

- *Conceptual understanding.*
- *Visualization.*
- *Mathematical reasoning.*
- *Mental mathematics* and *estimation.*
- *Aspects of thinking: thinking flexibly; thinking about thinking; algebraic thinking.*

Whenever possible strategies need to be considered that can *advance* students' *thinking*.

The aspects of *sense making* and *cognition* allow students to develop *mental mathematics strategies* for the basic facts and to create *personal strategies* for computational procedures.

Conceptual Understanding

Research shows that instructional settings can emphasize *conceptual understanding* without sacrificing skill proficiency.[3] An emphasis on skill proficiency and meaningless mastery will not contribute to any aspects of *conceptual understanding*.

Key indicators of *conceptual understanding* include the ability to:

- Talk and write about what has been learned in one's own words.

  During conversations and diagnostic interviews it becomes apparent that most students and adults lack this ability. Responses by subjects to requests for explanations of computational procedures are very similar. These explanations begin with,
  'First you write one number below the other. Then you start at the ones. ...'

  Some time ago, an instructor turned entertainer made a song out of the rote chants to find answers. The opening line of his song about teaching subtraction includes the key words of the chant familiar to most students,
  'You can't take ... from ..., so you go to the ... in the tens place.'

  Conceptual understanding implies that procedures and reasons for the procedures can be explained in one's own words, no matter how inadequate someone may judge these to be at the early stages of learning about a procedure. The acquisition of this ability is dependent upon students being given many opportunities to share explanations as part of their mathematics learning.

- Connect what has been learned to experiences outside the classroom and to other subject areas.

  Interviews yield many examples of students' inability to connect. Many students are unable to tell who might use the algorithmic procedures for multiplication or division and when and where these might be needed. Many are unable to connect decimal and integers to anything outside the classroom.

- Connect what has been learned to previously learned aspects of mathematics learning.

  This ability to connect and to tell how what is being learned is different and how it is the same from what was learned previously can facilitate understanding of new learning. For example, students who know how the divisive actions are the same and how they are different from the multiplicative action will be better able to connect the actions to their experience.

Visualizing

The ability to *visualize* is an important part of key aspects of sense making and the ability to solve problems. Fostering the development of *visual thinking* and the ability to form *visual images* requires special types of activities and special types of questioning. A provision and use of concrete materials, technology and visual representations is not sufficient. The right questions and proper problems need to be posed or presented as materials are handled. Without high-order thinking questions and open-ended problems little, if anything related to the desired mathematics learning outcomes will 'travel' from the hand or the hands, up the arm to the brain.

During diagnostic interviews many indicators of students' inability to *visualize* can surface. For example:

- Inability to visualize the numbers for given number names.
- Inability to illustrate or recognize the action of operations and the order of the numbers in pictorial representations.
- Inability to translate equations into pictorial representations.

A teacher's role in fostering the development of *visualization* is essential. Appropriate activities, problems and questions need to be considered during the planning stages. Without a conscious effort of attempting to focus on fostering *visualization* this important aspect of mathematics learning will not develop.

Mathematical Reasoning

The information processing or thinking that is part of learning and doing mathematics is not very different from most kinds of thinking that human beings do.[5] Greenwood[6] identifies seven criteria of mathematical thinking (p.144):

- *Everything you do in mathematics should make sense to you.*
  This sense making implies the presence of *conceptual understanding* which includes the ability to talk about mathematics in one's own words. Davis[5] reports what he calls 'disaster studies' since the subjects that were interviewed could not talk about the mathematics they had learned. This dilemma can be attributed to the type of requests that are made by some teachers, tutors or parents. For example:
  - *Just do what I tell you to do, it will work and you will get the correct answer.*
  - *Just follow these rules and steps.*
  - *Just memorize these steps and you will always get the right answer.*
  - *Just take this short cut because it is much faster.*
  - *Let me show you a trick that will always work.*

It is very discouraging to hear university students, and that includes teachers-to-be, utter comments like, '*Don't tell me why it works, just show me how to do it*' or, '*Just tell me what I need to know to pass the test.*' Many students at this level express gratitude to the teachers who told them what to memorize and who coached them to pass tests. It may be quite a challenge for these teachers-to-be to create settings of sense-making once they get a classroom of their own.

- *Whenever you get stuck, you should be able to use what you know to get yourself unstuck.*

  Sense making is a pre-requisite for this ability. Numbers and operations need to make sense before students can develop and acquire the *mental mathematics strategies* that will make it possible for them to get themselves unstuck.

  Students also need to be *confident* and *willing to take cognitive risks.* It is a little discouraging to observe students in classrooms and to interview students who lack these characteristics. They are afraid to try something that might be judged as incorrect.

- *You should be able to identify errors in answers, in the use of materials, and in thinking.*

  Sense making and *conceptual understanding* can be fostered in students by having them look at and examine examples as well as non-examples. Non-examples can become part of demonstrations as well as appropriate practice exercises. For example, one teacher concluded a lesson by recording calculation procedures on the chalkboard that contained errors and her challenge to the students was to try and find out what they thought her brain had been doing wrong. Practice sheets can be prepared which include descriptions of solution procedures that are incorrect or diagrams that are inappropriate or labelled incorrectly.

- *Whenever you do a computation you do a minimum of counting.*

  Perhaps the criterion could read, *do something other than counting*, since counting is not a strategy. *Mental mathematics strategies* and *personal strategies* are based on possessing a *sense of number* which includes *visualizing* and *flexible thinking* about numbers and do not depend on counting or counting by ones. The goal should be to do computations without counting.

- *You should be able to perform calculations with a minimum of rote pencil-paper computations.*

  It is suggested in the curriculum that students are given the opportunity to develop *personal strategies* for computation procedures. The development of a *personal strategy* ensures that there will be no rote pencil and paper computations.

- *When the strategy you are using isn't working, you should be willing to try another strategy.*

  This criterion is based on the assumptions that students are *confident* and *willing to take risks* and that they are able to *think flexibly.* If these characteristics have not been part of students' experiences while learning about mathematics, this aspect of *mathematical thinking* and reaching this goal presents a challenge.

- *You should be able to extend, or change, a problem situation by posing additional conditions or questions.*

  This ability can only be reached if it is included as part of learning mathematics whenever possible. Einstein stated that an imagination is more important than knowledge. Whenever possible, students should be faced with scenarios that ask them to consider responses to requests that begin with:
  'Pretend …'; 'What if …'; 'Let's assume that …'; 'What is another way of thinking about …?'; 'Who can think of a different way to …?'

There are many important specific learning outcomes that are related to or are part of the criteria for *mathematical thinking*. Reaching these outcomes is dependent upon careful planning and skilful actions.

Mental Mathematics and Estimation

Mental mathematics implies the use of *mental mathematics strategies* to arrive at the answer without the use of pencil and paper. The last part of this sentence seems redundant, but when students are asked during an interview to explain their strategy for calculating an answer 'in their heads' many will describe the procedure that is exactly the same as the one they were taught to use with a pencil and a piece of paper.

The ability to *visualize* numbers and *flexible thinking* about numbers are pre-requisites for *mental mathematics* activities. If, for example, students attempt to calculate the sum for two three-digit numerals in their heads, i.e., **259 + 647 = □**, they should experiment with different starting points or consider the digits in different orders. For example:

- hundreds – tens – ones;
- tens – hundreds – ones;
- ones – hundreds – tens;

and then comment on their preference and why that is the case. Does everyone prefer the same strategy?

Estimation of number implies the use of a *referent* or a *benchmark* as a prediction is made for the answer to, *About how many*? For example, the number of fingers on both hands, or ten, is used as a *referent* to arrive at an estimate reported as, *About _ tens*, for the number of counters displayed on an overhead projector. One-half, or **0.5**, can be used as a *benchmark* for placing given decimals on a number line.

Estimation can involve predicting answers or checking the reasonableness of answers for computational procedures. At one time students were taught rules for rounding numerals that became part of a rote procedure to make predictions about answers. Since *number sense* develops gradually, the assumption that all students should use the same numerals as part of their estimation strategies does not make sense. Students should be allowed to choose the numerals they are able to manipulate mentally, or their 'nice numbers', as they are referred to in some classrooms. Five or six different students could use different 'nice numbers' for their estimates for **642 ÷ 7 = □**.
For example:

700 ÷ 7   600 ÷10   650 ÷ 10   630 ÷ 7   650 ÷ 10

Communicating about the teaching of mathematics not only requires the use of correct language, but also specificity. Without specificity it will be difficult, if not impossible, to implement what authors had in mind about teaching, learning and assessment. During discussions about aspects of teaching, statements are often uttered that are of a very general and subjective nature. The references teachers use also include statements that are much too general, and that is true of documents from Ministries of Education, including the mathematics curriculum.(7) For example, what does the statement, '*demonstrate fluency with mental mathematics and estimation*' (p.6) mean? How might different readers of this statement answer the following questions:

- How is *fluency of mental mathematics and estimation* defined?
- How is *fluency* assessed?
- What are some possible criteria for *fluency*?
- What are key indicators of *fluency*?
- What are indicators of lack of *fluency*?

Teachers-to-be as well as teachers can be heard to make comments that include terminology like '*clear understanding*' or '*until the students are comfortable*' as goals of teaching are discussed. There is not one listener who is able to guess what the '*clear*' might refer to or how '*degrees of comfort*' can or might be assessed.

Statements about assessment of ability to estimate often make reference to '*reasonable*' estimates without telling a reader whether '*reasonable*' is determined according to a student's or a teacher's notion about this ability or how *reasonable* might be defined.

General expressions and terms in references require definitions, criteria or examples. Lack of specificity can be a roadblock for attempts to accommodate the *critical components* of the curriculum. Making mathematics meaningful for students requires that statements about teaching, learning and assessment are clear and are not open to different interpretations, misinterpretations, or are impossible to interpret.

Aspects of Thinking

The mathematics classroom is an ideal setting to foster different aspects of thinking since many opportunities exist to pose *high-order thinking questions*. For example, requests can be made to provide reasons for any moves that are made or to justify any responses that are elicited. On the other hand, it is also easy to create settings that do not include any *high-order thinking questions*. Such settings focus on: brief answers; responding quickly; mastery learning, however that may be defined; rote pencil and paper calculations.

Many topics and ideas discussed in a mathematics classroom are suitable to foster *flexible thinking*. A well-phrased and well-placed request can lead students to change or to employ new strategies.(8) For example,

- *What might be another way of doing this?*
- *Try to use at least two different ways of solving the problelm.*
- *After you have listened to others explain how they solved the problem, try one or two ways that are new to you.*

A positive response to requests like these requires *thinking about thinking*. This opportunity to *think about their thinking* also exists while editing what has been written or when instructions for riddles or games that have been invented are modified.

Any documentation of responses that is collected, kept and returned on a later date can be used to *advance thinking*. An appropriate request for this purpose is required.

For example:

- The challenge is presented to modify a sketch that has been prepared according to new specifications, i.e., Now try to re-design your flag to show two patterns that differ.
- After a discussion the request is made to use the information from the discussions to modify a definition that has been created for a term, i.e., fraction.
- The request is made to think of ways to modify answers that were calculated, i.e., What two different things could you do to double each of the answers?

One aspect of mathematical thinking is labelled *algebraic thinking. Algebraic thinking* is not a separate component of mathematics learning, it is part of ongoing mathematics learning.
For example, aspects of *algebraic thinking* are used when:

- A hidden member or hidden members of growing or repeating patterns are identified by looking at the members of the patterns that are not hidden and using these to make generalizations about the patterns.

- A box or a letter is used in an equation, i.e., **3 + ☐ = 8** or, **3 + n = 8**, as a number sorter as attempts are made to record an equation that is true.

- Generating rules for a property, i.e., *'Every number added to zero, or zero added to any number will always be the other number'*; expressing it in general terms, i.e., **n + 0 = n** and **0 + n = n**, and then testing the rule.

- Figuring out and describing what a 'magic calculator' does to each of the numerals entered and describing what is happening. For example, when **3** is changed to **7; 4** is changed to **9**; **5** is changed to **11**; etc., students conclude, *'Double it and add one'* or, *'Twice the number plus one'* or, '**2** *times n then add* **1**, or **2n + 1**.'

- Expressing relationships by using formulas. For example, *'I can find the area of a rectangle by counting the squares along the base, then counting the number of rows of squares along the height and then multiplying these numbers to find the area'* or, *'length of the base times length of the height'* or, 'base *times height'* or, '**b x h = area**' or, '**b x h = a**.'

Other aspects that are considered part of algebraic thinking include such ideas as:

- Order of the operations.
- The commutative (**a + b = b + a**), associative [**(a + b) + c = a + (b + c)**] and distributive properties [**a(b + c) = ab + ac**].
- Graphing to illustrate properties and to explain change.

Aspects Related to the Goals for Students

An environment needs to be created which continues to contribute to fostering *characteristics* in students that include:

- *Confidence* in mathematics.
- *Willingness to take risks.*
- Ability to *communicate mathematically*; orally and in written form.
- Ability to *connect* to previous learning, ongoing learning; future learning and to experiences outside the mathematics classroom.
- Exhibiting *curiosity*.
- Use of *imagination*.

Whenever possible, *tolerance for ambiguity* as part of mathematics learning should be recognized and acknowledged.

Confidence

According to the curriculum[2], *Mathematics education must prepare students to use mathematics confidently to solve problems.* (p.4). The development of confidence can present a challenge. Previous experiences from in and out of classrooms may have resulted in a lack of confidence in mathematics and a negative attitude toward the subject even at this early stage of learning. It could also be possible that any strategies that are employed to build confidence in a classroom differ from what takes place or is talked about in the home.

Many adults still believe that an important part of 'knowing mathematics' means being either right or wrong, and 'being good at mathematics' means or may mean some or all of the following:

- The ability to give correct answers quickly.
- Knowledge of a fastest way to 'do' things like using calculating procedures
- Knowledge of a best way to solve problems.
- The ability to calculate answers quickly in one's head.
- The ability to recite memorized rules, procedures and formulas.
- Receiving a high score on a timed test of rote computations.

There are adults who will share not only with their children, but with other listeners as well, that they did not like mathematics, found it difficult and were not good at it. This type of background experience will impact students' disposition toward mathematics learning and present additional challenges for teachers.

Fostering and maintaining *confidence* in mathematics is very dependent upon how mathematics is taught and how mathematics learning is assessed. Strategies that can contribute to building confidence in mathematics can include:

- Providing open-ended tasks and problems which will lead to the conclusion that there are different procedures and ways that can be used to arrive at the desired results and conclusions.
- Allowing for the use of personal or natural language to explain ideas and procedures.
- Allowing for sufficient time to complete tasks rather than placing an emphasis on speed.
- Giving credit and recognition to explanations or parts of explanations, rather than just a conclusion or correct answer.
- Basing assessment and marks only on tasks that have been completed and/or attempted.
- Using questions of the type that let students know that whatever is stated as part of a response is accepted.
- Designing and using appropriate or meaningful assessment items.

Willingness to Take Risks

Almost all classrooms include students who are reluctant or unwilling to take cognitive risks even as early as in grade one. These students frequently make requests like, '*What am I to do*?' and, '*How am I to do this?*' They seem to be afraid that if they try something on their own, they might make a mistake; they want to make sure that whatever they do is correct. It would be ideal if it was possible to say to these students, '*Just try something and let's see what happens*' to get them to take a risk, but that does not work. Evidence exists that shows open-ended instructions and activities encourage risk taking in students.[8] In these settings students learn to realize, and for some this is a gradual process, that there is not just one correct way of 'doing things' in mathematics and that it is acceptable to try and to experiment with different procedures.[9]

Teachers play a key role in fostering the characteristic of *willingness to take risks*. Two important components of this role include early recognition of any *risk taking* and the accommodation of all types of responses. An acknowledgement and praise of any signs of risk taking can lead to further risk taking which in turn can build confidence to make this part of one's behaviour. In this type of setting it is essential that students learn that all of their responses are accepted, valued and in some way accommodated. This acceptance may present a small challenge for teachers, but the greater challenge may lie in preparing all classmates to do the same, no matter how creative, humorous, imaginative, deviant or unusual the responses from some students may be.

Ability to Communicate Mathematically

One of the guidelines from research that can be implemented with confidence is that 'students need opportunities to engage directly in the kind of mathematics that they are to learn.'[3] Since the ability to *communicate mathematically* is a goal of the mathematics curriculum, opportunities to talk and write in the mathematics classroom need to be provided. As students talk or share what they have written and as their classmates react, ideas can get clarified, modified or reinforced. *Generalizing ideas through communication is vital when building mathematical language.*[10]

In the book *Getting into the Mathematics Conversation – Valuing Communication in Mathematics Classroom*[11] the authors develop and present a theoretical framework that places the aspects of *communicating*, which include reading, writing, listening and speaking, in the realm of pedagogical content knowledge. These aspects of discourse are part of the learning outcomes of the mathematics curriculum and they are an important component of this book. However, this theoretical framework is beyond the goals and the scope for this book.

The provision of opportunities to talk can be accommodated in cooperative settings where procedures, ideas and strategies are discussed with a partner. Appropriately placed and phrased questions or requests can result in discussions that involve thinking and *thinking about thinking*. More talking is involved when groups are asked to make decisions about what to report to the whole group and which parts of their discussions should be included. The results of the reports can foster *flexible thinking* if students are challenged to try to use some of the new ideas or approaches they have listened to. Evaluative skills are developed if groups are requested to compare reports and to draw conclusions.

A request to get students to write something about some aspect of mathematics learning has been met on more than one occasion with, '*We don't do that in mathematics.*' Many students may believe that writing is not an important part of learning mathematics; it does not lead to any new learning; and it is not an enjoyable task.[12] However, quite a few students changed their minds after writing tasks were made part of many of their activities in mathematics throughout the year. Many important learning outcomes can be met in a setting where students get opportunities to write, to report and to discuss and compare what they have written. Settings that make talking and writing an integral part of mathematics teaching and learning contribute to students' language development, reading comprehension and the development of evaluative skills.

Connecting

The ability to *connect* can be thought of as a problem solving strategy. The opportunities that exist to *connect* while teaching about mathematics need to be taken advantage of. For example:

- Connecting to experiences outside the classroom:
  Who uses decimals? When? Where? Why? and, *Who adds decimals? When? Where? and Why?*

- Connecting to previous learning:
  How is multiplication different from what we have learned before? How is it in some way the same?

- Connecting to ongoing learning:
  Why do you think we need to know something about integers?

Curiosity

A student's curiosity is dependent upon a level of *confidence* and *willingness to take risks*. These are pre-requisites for being willing to ask questions of the type,
'*Why?*'; '*What if …?*'; '*What else could happen?*'; '*I wonder what would happen …?*'

Use of Imagination

It was Einstein who said that *an imagination is more important than knowledge*. There exist opportunities while teaching about mathematics when it may be possible to make requests of the following type:

- *Try to think of a different way or a way nobody in this classroom has thought of.*
- *How do you think people from a different planet might try to solve this problem?*
- *How might you solve this problem if you were to do everything wrong?*
- *How could you teach someone to calculate the answer if that person could not talk?*
- *How could you teach someone to calculate the answer if that person could not hear and could not use pencil and paper?*

General Issues Related to Aspects of Assessment and Reporting

The assessment of aspects of *sense making*, aspects of *cognition* and the *characteristics* that are to be fostered in students requires procedures that differ from the timed tests consisting of rote computational procedures or simple multiple choice items. A variety of techniques have to be employed that will provide information about indicators of the important aspects of mathematics learning. The data that are collected have to be translated into statements for reports that use specific and non-subjective language enabling everyone who reads such statements to interpret them in the same way.

Assessment

It is not possible to talk about the teaching and learning about mathematics without talking about aspects of assessment. A few issues of a general nature are described in the paragraphs that follow. Specific examples for suggestions to collect assessment data and reporting to parents are included for each topic.

When the focus of teaching mathematics is on mastery of paper and pencil computational skills, timed tests will yield results that can be reported in terms of percentages and grades can be assigned to different ranges of these percentages. Such a procedure seems straight forward and objective. However, a closer examination will point to at least two areas of possible difficulty.

Definitions of mastery that are adopted may differ. Some may aim for a **90%** *level of accuracy* or even higher, while others may adopt an **80%** *level of accuracy*. How decisions about adopted levels of mastery are made can result in some fascinating discussions. Different schools or different areas may also consider different ranges of percentages for grade conversions, another possibility for arbitrary or subjective decision making.

Assessment results that focus on rote procedural learning of mathematics can result in high grades for many students. Should these students enter university they are likely to find out these high grades do not imply the presence of any conceptual understanding of mathematics. From time to time this dilemma surfaces in newspapers reports or in letters to the editor when people describe how the high grades they received incorrectly led them to believe that they were numerate and knew a lot about mathematics.

The learning outcomes and goals of the mathematics curriculum require a range of assessment strategies. Assessment data collected about many aspects of *sense making*, *aspects of cognition* and the *characteristics* to be fostered in students cannot be translated into percentages. If criteria listed for these components of learning are to be translated into grades for students, care needs to be taken that none of the entries in an assessment rubric are subjective since most statements of this type are impossible to interpret. A close and critical inspection of many checklists that are recommended for teachers will indicate that many of the entries are subjective.[13]

Informing parents about achievement related to the *critical components* and the *goals for students* necessitates taking advantage of opportunities to collect assessment data in situations that are embedded in teaching-learning settings. For example, many valuable pieces of information can be noted as students:

- Talk with a partner or in a group.
- Report what has been talked about in a group.
- Read for their classmates what they have written about a topic.

- Share their strategies of solving a problem.
- Share their strategies to get unstuck.
- Explain and illustrate their personal computational strategies to classmates.
- Explain what might be wrong and why they think that an error has been made.
- Explain and illustrate their thinking with objects.
- Explain how a sketch they have prepared shows what they are or were thinking.
- Examine examples that: illustrate incorrect thinking; include incorrect use of materials; show inappropriate sketches; and/or show incorrect calculations for solutions and explain what the errors are and why they may have been made.
- Share their rules for a game they have invented and then try to get their classmates to play the game.
- Explain who would use the mathematics they have learned and where and when.
- Share problems they have authored or co-authored.
- Orchestrate discussions as classmates solve riddles they have created.
- Explain what they think the missing parts for a data plot might be.
- Explain their estimation strategy for tasks involving number, computation or measurement.
- Explain their speculations about tasks that elicit several responses to requests like:

  Which one do you think is different or does not belong?

  What do you think could come next?

Specific examples for the entries in the above list as well as for other tasks related to appropriate or meaningful practice are described in the book. These examples will make it possible to author report statements that provide readers with indicators of meaningful mathematics learning.

It is possible to construct tests items that assess aspects of cognition. However, in order to be fair, most items of this type require detailed instructions and opportunities to explain thinking. Many test settings include inappropriate questions and opportunities to explain reasons for a response are not given. To illustrate this point, consider an item that shows a repeating pattern with the request, *What comes next?*

Since repeating patterns can be extended in many different ways and they can be changed to increasing patterns, it is inappropriate and unfair to ask a question in this simple form and to have one answer that is considered correct. Any responses that differ are marked wrong and in turn lead the recipients to conclude lack of knowledge about a pattern or about patterns. Such a conclusion has a high probability of being wrong. If the intent is to elicit one specific response very detailed instructions are required. Examples related to this scenario are discussed in *Making Mathematics Meaningful in the Primary Grades*[14] and some are included in this book.

Published tests also include items that are unfair, inappropriate and do not provide any meaningful knowledge about mathematical thinking. These tests may also include content that is mathematically incorrect.[15] There are times when items appear on such tests and are retained because they meet criteria of those who design tests rather than meeting aspects of mathematical thinking. During a meeting the question arose about a test item that asked students to find the perimeter of a figure with an irregular shape that was labelled with different units. The concern was addressed with, '*This is a two-step problem that assesses understanding or perimeter.*' This response implies that any student who solves the item incorrectly does not understand how to solve two-step problems and does not 'understand perimeter.' Neither of the conclusions may be true.

The *critical components* and the *goals for students* that are part of the framework of the mathematics curriculum require that some of the assessment procedures that have been used be re-examined. The examples and discussion are intended to illustrate that the assessment of many learning outcomes of the mathematics curriculum requires a careful examination of the items included on tests. There exists a need for types of assessment with items that cannot be simply classified as right or wrong. Assessment data need to be collected to make it possible to write reports for parents that share meaningful information about the *critical components* and the important *goals* for students.

Reporting

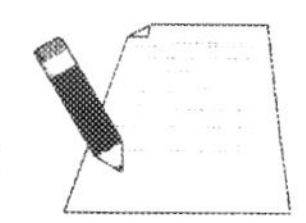

It is essential that parents or other stakeholders are informed about any progress that has been observed and noted with respect to the *critical components, goals for students* and the *major learning outcomes* of the mathematics curriculum. That means information with respect to key aspects of *sense making* and *cognition* needs to be shared as well as information related to any indicators that have been collected about *characteristics* in students that are fostered in the mathematics classroom.

The statements that are shared need to be specific or non-subjective. This emphasis on specificity may seem exaggerated, or trivial, but it is all too easy and common to generate statements about mathematics learning that may sound impressive, but are meaningless. For example, what questions should and could be asked of the authors of the following examples?

- *Has mastered the nine times table.*
- *Has learned to divide three-digit numbers.*[16]
- *Johnny needs help with fractions.*
- A category titled: *Meets Expectations* is described as, *'The work satisfies the most basic requirements of the task, but is flawed in some way.'*[16]

Many references about mathematics assessment used by teachers include terms and phrases that require definitions, examples or criteria before these statements can be of any practical use. At first glance, some of these may seem appropriate but closer inspection shows that these statements are impossible to interpret. What specific skills and ideas might the authors who reached the following conclusions about students they interviewed have had in mind?

- *Has good number sense.*
- *Lacks number sense.*
- *Lacks spatial sense.*
- *Makes reasonable estimates.*
- *Does not understand fractions.*
- *Knows the basic multiplication facts.*
- *Has mastered the basic division facts.*
- *Is a good problem solver.*
- *Has conceptual understanding.*
- *Is unable to visualize.*
- *Is able to think flexibly.*
- *Is confident.*
- *Can communicate mathematically.*

One Special Education Teacher shared the following about a student that was sent to her for assistance, '*Does not understand Chapter 4.*' It is a challenge to try and think of a statement that is more general than that! Two very challenging tasks exist. Meaningful assessment data need to be collected and then translated into specific statements for report cards, parent - teacher interviews or for anyone who may offer learning assistance.

General Issues Related to Aspects of Diagnosis and Intervention

There exist many possible reasons for teachers in the intermediate grades to be prepared to administer diagnostic procedures and for the need to prepare effective **IEPs**. Some of these reasons might include:

- *Sense making* is a gradual process and some students may not have acquired the necessary prerequisite skills and ideas for the mathematics they are to learn.

- Students may have learned their mathematics in settings that did not focus on the *critical components* and *goals for students* of the mathematics curriculum. Some teachers '*tend to feel more comfortable with and capable of teaching lower level knowledge and skills rather than more complex processes*' (p.76). [17]

- Students may have learned their mathematics in settings with an emphasis on rote procedural learning and performing well on timed tests rather than on the development of *conceptual understanding*.

- Some students may come from home environment where mathematics learning is thought of as 'there is a best way' and all it takes to be 'good' in mathematics is 'more practice' to 'recall facts' or to be able to 'perform arithmetical calculations quickly.' Such a background can be detrimental to efforts in classroom settings that attempt to accommodate the *critical components* and the goals of the mathematics curriculum.

Diagnosis

Just as it is unthinkable for a medical doctor to draw diagnostic conclusions about patients without talking to them, it would be just as impossible to make evaluative comments about the different aspects of mathematics learning without having a conversation or conducting an interview. Without knowing how students think it is not possible to plan effective **IEPs**. Responses on written tests do not provide the information that is required for effective intervention.

Probing thinking is research at a high level; the teacher takes the role of a cognitive diagnostician. Well-phrased and well-placed questions that are free of mathematical terminology are required. Care needs to be taken not to contaminate or adulterate the dialogue during an interview. In a one-on-one setting all types of responses have to be accommodated.

One of the greatest advantages of an interview over a written test is the fact that various adjustments are possible during an interview. Rudnitsky et.al.,[18] describe possible ways of locating a proper level of difficulty and suggest four possible adjustments or strategies: *illustration*, *redirection*, *particularization*, and *generalization*.

- *Illustration* involves asking subjects to represent or define an idea, operation or procedure in terms more concrete than those used in an earlier response. Requests for responses could include illustrating thinking with objects, with a sketch or with a word problem.

- *Redirection* entails changing a line of questioning to either a simpler idea or task or a more difficult one, depending on a subject's responses.

- *Particularization* is employed when an error is made or a generalization is inappropriate. The intent of the questioning sequence, which can be thought of as several redirections to simpler ideas or examples, is to determine whether or not a subject is able to realize that the error or the incorrect generalization leads to a contradiction. For example, what answer would a subject who records **46** as the answer for **52 – 18 =** ☐ record for **12 – 8 =** ☐? Recording **16** and suggesting that this is the answer yields valuable diagnostic information. On the other hand, a lot is learned about a subject's insight and ability to think about thinking, if the subject suggests that a mistake has been made and this is followed by a request to change the answer for the first equation.

- *Generalization* comprises an attempt to elicit a rule for specific types of computations and attempting to determine whether or not a subject is able to apply the rule to different examples.

- The interview strategy of *clarification* simply consists of a request to have the subject talk further about an idea, sketch or computation.

Although these strategies are essential to probe thinking during diagnostic interviews, they can also become a valuable part of the dialogue that takes place while teaching about mathematics.

Diagnosis may be required when a new student transfers into the classroom or at the beginning of the year when some students may be encountered who find it difficult to cope with the mathematics that is presented.

At times it may be tempting to make a guess about the possible reason or reasons for the errors committed by students and to have a plan for intervention in mind. Research indicates that those guesses can be wrong more than **50%** of the time and point to the necessity of follow up interviews in order to plan appropriate intervention.(19) The following are three examples from files of interview transcripts in support of these findings.

- For **32 – 17 =** ☐ the student recorded **25** as an answer. When asked to explain how he arrived at the answer, he stated, '*Two minus seven I can't do, so twelve minus seven is equal to five and thirty take away ten is equal to twenty.*'

- The student's paper showed five examples solved as follows:

$$\begin{array}{r} 26 \\ \times\ 7 \\ \hline 1442 \end{array} \qquad \begin{array}{r} 42 \\ \times\ 6 \\ \hline 2412 \end{array}$$

  When asked to explain how he got the answers he stated, '*Six times seven is equal to forty two, record the two and carry the four tens. Seven times twenty is one hundred forty tens plus the four tens is equal to one hundred forty-four tens.*'

  The second example was explained in the same way, '*Six times two is twelve, record the two and carry the one ten. Six times four tens is equal to two hundred forty tens plus one more ten equals two hundred forty-one tens.*'

- After having read an article about the importance of interviews to help students with subtraction,[20] one mother from New York sent a letter to the university with the following anecdote about her son. The items on an activity sheet with fifteen subtraction items that did not involve renaming had been marked wrong. The boy had solved the problems in the following way:

$$\begin{array}{r} 42 \\ -\ 11 \\ \hline 33 \end{array} \qquad \begin{array}{r} 65 \\ -\ 24 \\ \hline 49 \end{array} \qquad \ldots$$

  For homework the teacher had assigned ten similar items with the simple request, '*Think.*' When the mother asked the boy about the items, he responded with, '*But Mom, I am thinking. When there is not a sign in front of the numbers, I add and when there is a sign there I do what the sign tells me to do.*'

The responses by students during conversations and interviews provide specific information about what is required during intervention. Without this information it is not possible to plan appropriate **IEP**s and it is likely that action that is taken will be inappropriate.

Throughout the book pre-requisite skills, procedures and ideas are identified that can be adapted for diagnostic purposes.

Intervention

At one time students who did experience difficulties or in some way did not keep up with the others were sent out of the classroom for what was called remediation. For many of these students this remediation consisted of skill oriented practice. If during this absence from the classroom, the instruction focused on *sense making* and *conceptual understanding*, the students who left the classroom would have fallen further behind in their learning about mathematics.

There are some people who will share the beliefs that mathematics learning is like learning to play the piano or like learning how to play basketball, which means extended practice is essential and eventually this practice can lead to fostering understanding. Both parts of such beliefs are inappropriate as far as reaching the goals of the mathematics curriculum. Students in the upper intermediate grades who require intervention may have had many experiences with practice settings. More of the same will not be of any help. Commercial institutions that focus on skill development and on the preparation to write tests authored and used by these institutions will not be of any long term benefit to students. Intervention settings need to focus on key aspects of *sense making* and *cognition*: *number sense*; *conceptual understanding*; *visualization* and *mathematical reasoning*. It is very likely that students who are recommended for intervention need to gain *confidence* in mathematics.

Effective intervention **IEP**s require specific learning outcomes and specific language.[21] That means that any diagnostic data collected from observations, listening to reports, assignments of appropriate practice settings, and/or responses collected during interviews have to be translated into specific learning outcomes that relate to the key aspects of *sense making* and *cognition*. These types of learning outcomes will make it possible to measure progress or growth.

Students who require intervention need to be provided with appropriate practice. This type of practice needs to focus on aspects of *sense making* and *conceptual understanding*. Examples of appropriate practice settings are illustrated in the book.

The challenges that are part of planning for the many essential components of mathematics learning as part of teaching, assessment, reporting, diagnosis and intervention are indicative of the complexity of the task. The importance of the key role of a skilful teacher is illustrated and it clearly becomes evident that this role cannot be played by a book or a computer.

For Reflection

1) What questions would you ask a person who during a discussion states, '*I recognize fluency with mental mathematics or estimation when I see it.*'? Why would you ask these questions?

2) Many people and teachers believe that memorization is an important part of mathematics learning. What skills, procedures and ideas do you think these people would include in a list of things that need to be memorized? Compare your list with a list generated by someone else. How might these people think the memorization they suggest is accomplished? What possible benefits might be attributed to the act of memorizing or the results of memorization?

3) When some people talk about memorization, they use the descriptor rote. Do you think this descriptor is needed? Why or why not?

4) Many educators agree with educators and authors who state that rote learning is an oxymoron. Two co-authors[22] list thirty-four advantages of learning with understanding and the same number of disadvantages for rote learning. What do you think are five important advantages of learning with understanding? Why are they important? What five possible disadvantages of rote learning might you share with someone?

5) What would you say to someone who shares the conclusion *Mathematics is and always has been a matter of being right or wrong. I was good at mathematics. I always got high grades. I got the right answers and I knew how to get them quickly*?

6) A first year teacher is assigned to a group of grade four students. Early in the year the teacher expresses amazement and disbelief about the wide range of abilities with respect to all aspects of mathematics learning in the classroom. The teacher is at a loss for explanations and words and is looking for some answers. What possible reasons for this divergence in abilities would you include in your request for a possible explanation for this dilemma?

7) A newspaper article about mathematics learning[23] included the comment from the book *How Mathematicians Think: Using Ambiguity, Contradiction and Paradox to Creative Math* by B. Byers, '*math anxiety often has more to do with inept teaching and bad experiences in grade school than from innate inabilities*' (D9). What are some possible examples of inept teaching and bad experiences that Byers might have in mind?
How might Byers respond to King[24] who concludes, '*Attributing teaching and learning failure to something called "math anxiety" serves no purpose except to provide a built-in excuse for inadequate performance on both sides*' (p.127)?

8) What questions would you ask of the author who included the following on a lesson plan, *Once all students feel comfortable, I will go on to the next idea,* and *I want the students to have a clear understanding of what I am teaching?*

9) What questions would you ask of an author who talks about *flawless mastery*?

Chapter 2 – Developing Number Sense
Whole Numbers, Fractions, Decimals, Percent, Integers

The Importance of Number Sense

Number sense is a key aspect of *sense making* for ongoing and future learning of mathematics. In the intermediate grades *number sense* is a pre- and co-requisite for enabling students to reach the important learning outcomes of the mathematics curriculum related to:

- *Mental mathematics strategies* for the basic multiplication and basic division facts.
- *Personal strategies* for the four operations (**+** , **-** , **x** , **÷**) for whole numbers and decimals.
- *Personal strategies* for the addition and subtraction of integers.
- *Estimation strategies* related to number: whole numbers; decimals; fractions and percent.
- *Estimation strategies* related to computational procedures for whole numbers and decimals.
- *Mental mathematics* with whole numbers and decimals.

The strategies that are part of the development of *number sense* contribute to *conceptual understanding* and enhance transfer to new learning about mathematics. Making sense of numbers is a capability of students who are numerically powerful.[1]

Number sense is the key foundation for *numeracy*.

Components of Number Sense

The key components of *number sense* include *visualizing*, *flexible thinking* about numbers, *relating* numbers and numerals, *estimating* number, *connecting*, mental calculations or *mental mathematics*, and aspects of *recognition*.

Visualizing

When students hear the names for numbers, i.e., *nine hundred sixty five*; *three thousand two hundred two*; *four-tenths*; *sixty four hundredths*; *two-thirds* or they see these number names or numerals in print, i.e., **965**; **3 202**; **0.4**; **0.64**; **2/3,** they should be able to *visualize* the numbers for these names.
The ability to *visualize* numbers enables students to:

- *Estimate numbers.*
- Make predictions about the answers for calculations.
- Make comments about the reasonableness of answers.
- Develop *mental mathematics strategies.*
- Develop personal computational strategies.
- Do *mental mathematics.*

Flexible Thinking about Numbers

The realization that a number can be represented in different ways and thus have different names is important for the development of *mental mathematics strategies* for the basic facts; for inventing *personal strategies* for computational procedures and for *mental mathematics.*

The realization that different numbers of hundreds, tens or ones can be used for the same number is an indicator of being able to *think flexibly* about numbers.
For example, **265** could be:

- two hundreds, five tens and fifteen ones or,
- one hundred, sixteen tens and five ones.

Writing a numeral in expanded form, i.e., **200** + **60** + **5**, or listing each of the digits in a numeral and stating beside each digit the place value position of the digit, i.e., **2(100)** + **6(10)** + **5**, are not different names for a numeral and therefore are not indicators of being able to think flexibly about numbers.

Relating

The ability to *relate* includes comparing and ordering numbers and numerals and to use the appropriate mathematical language to describe the results of such activities. The mathematical language of *relating* numbers or numerals can include such terms or phrases as: *is greater than*; *is less than*; *is between*; *is close to*; *more*; *fewer*, *as many*; *the same number of*, *about the same*; *even*; *odd*; *square*; *triangular*, *digits*; *sum* or *difference of digits*; *multiple*; *divisible*; *factor;* *prime*; *composite*.

Becoming conversant with the language of *relating* implies that students need to be given many opportunities to talk and write about numbers and numerals.

Estimating

Anyone who does crossword puzzles knows that the terms guess and *estimate* mean the same thing to many people, at least to some who design these puzzles and those who try them. In the mathematics classroom, *estimation* refers to something very different from guessing, and students need to know the difference between requests that ask for one or the other. Confusion needs to be avoided by resisting the temptation of adding descriptors to one or both of these requests. For example, requests or acknowledgements that include the descriptors 'good' or 'intelligent' for guesses and 'reasonable' or 'logical' for estimates are inappropriate because they are subjective. One term that must be at the top of a list that can create confusion is when someone talks about *guesstimates*.

Estimation is a cognitive process and it is dependent upon a person's number sense. *Estimation* of number involves making use of known information such as a *referent* or a *benchmark*, and arriving at answers for such questions as:

- *About how many tens* (hundreds: etc,) *are there?*
- *Where would the numeral appear on a partially labelled number line?*

Every response to a request to estimate which involved the use of a *referent* or a *benchmark* has to be considered as being logical. However, responses will differ since the development of *number sense* is a gradual cognitive process and is unique to every individual.

Connecting

The ability to *connect* what is being learned to one's experience is an important indicator of sense making. It is not too difficult to connect number and numerals in the tens and hundreds to events or actions from students' experiences. It is a challenge to have students attempt to visualize and connect as many key numbers and numerals as possible to groups of objects, animals or people that may be encountered in their experience, i.e.,

1 000; **10 000**; **100 000**; **1 000 000** or certain multiples of these numerals.

Mental Calculations

The ability to derive and employ personal strategies for calculating solutions without the use of pencil and paper is dependent upon the ability to *visualize* numbers and to *think* about numbers *flexibly*. Without these abilities, strategies devised and shared by someone else will have to be adopted in a rote procedural manner.

Recognizing

Experiences with materials that clearly illustrate the properties of the numeration system will foster the ability to report an initial reaction for the request, *What number do you think is shown?* when a number is displayed. This type of a response should not be based on any counting but it may involve the use of a *referent*. Reactions may include comments of a very general nature, but these responses can be indicative of *confidence* and *willingness to take risks*. For example, after briefly looking at a picture showing a group of people, the responses could be in the form of:

- *The group reminds me of the number _ .*
- *I think there seem to be more than _ , because it reminds me of ....*
- *I think the number of people is less than _ , because I am thinking of ....*

Any explanations that accompany these types of responses can yield valuable information about indicators of *number sense*.

Number Sense: The Importance of the Role of a Teacher

The components of *number sense* that have been identified and described are in part supportive of the suggestion that it is valuable to view many aspects of *number sense* as a by-product of learning about mathematics rather than direct teaching.[2] A *Number Sense Focus Issue* of the *Arithmetic Teacher*[3] includes various pointers by different authors for teaching strategies and classroom settings deemed conducive to fostering the development of *number sense*. The suggestions made by different authors include:

- *Number sense develops gradually as a result of exploring numbers, visualizing them in a variety of contexts and relating them in ways that are not limited by traditional algorithms* and *doing mathematics in an environment that fosters curiosity and exploration* (p.11).

- *Exposing children to messier aspects of everyday problem solving and placing more emphasis on thinking about various procedures that can be used to solve a problem and on interpreting the answer that these procedures produce* (p.16).

- *Promoting techniques and planning for the development of number sense for all parts or throughout a lesson* (pp.19 and 21).

- *Having students discuss the understanding of the 'hows' and 'whys' of number* (p.25).

- *Creating a climate that encourages pupils to ask 'why'* (p.29).

- *Selecting activities that give students the opportunity to verbalize relationships that demonstrate the acquisition of good number sense* (p.38).

- *Encouraging students to work on tasks in groups of three or four, relating the operations of arithmetic to real-world models, making children aware that not every mathematics problem has a single answer, challenging students to make generalizations* (p.46).

- *Provoking each pupil into constructing his or her own knowledge of numbers and the relations among them* (p.50).

- *Selecting appropriate activities conducive to the development of number sense* (p.55).

Some of the terms included in the quotes require examples since they are of a general nature or subjective. However, generally speaking, the recommendations point to the importance of the role of a teacher. Effective classroom settings need to be created, appropriate activities for an individual and for groups need to be selected or designed and small group discussions as well as whole group discussions need to be orchestrated. Selected examples for effective settings are discussed in this chapter as well as in parts of other chapters. It will be shown how the presence of *number sense* facilitates learning the *basic facts*. Examples that entice use of imagination are discussed. Problem settings that provide opportunities to share thinking are described.

The preparations and orchestrations required for fostering the development of number sense are similar to classroom practices that promote *numerical power*.[1] According to Charles and Lobato these practices include:

- Students having sufficient opportunities to develop meaning of numbers and to explore numerical relationships before being asked to learn about the basic facts and computational procedures.

- Promoting creativity by encouraging multiple solution approaches.

- Providing opportunities to demonstrate numerical reasoning not only during classroom activities but also as part of assessment, practice, and homework.

- Encouraging students to develop autonomy by using reasoning as verification rather than relying on a teacher for correct answers (p.41).

The pointers for effective classroom settings required to foster *number sense* and *numerical power* described by the different authors also make it possible to imagine and to describe classroom settings that would not be conducive to fostering the development of these key aspects of mathematics learning.

The complexity of creating an effective classroom atmosphere reinforces the conclusion reached by Reys[4], *'There is not now, never has been, and it is hoped never will be a genuine substitute for a good teacher who knows how and what children need to learn and when they need to learn it.'*

Number Sense: To Ten Thousand and Beyond

By the end of grade 3, student activities related to the development of *number sense* take the students to **1 000**. In grade four the teaching about numbers goes to **10 000**; in grade five to **1 000 000**; and in grade six understanding of place value is extended beyond one million.

Key Ideas from the Previous Grades

There exist two main reasons for a brief description of the key ideas for the components of *number sense* from the primary grades. If students find it difficult to cope with an ongoing program in the intermediate grades, it could be that these students lack key pre-requisites as far as *number sense* is concerned. Sample tasks from the ideas that are described can be selected for a diagnostic conversation or a diagnostic interview with these students. The second reason has simply to do with becoming aware of some of the main ideas that are part of mathematics teaching and learning in the primary grades.

There could be several reasons for students lacking some of the pre-requisite ideas and skills related to the development of *number sense*. *Number sense* is part of the conceptual domain and it develops gradually. Some students may have missed an important part of an instructional sequence. There may be students who come from classroom settings that did not focus on fostering the development of *number sense*.

Rational Counting

When presented with any ordered sequence of numbers or numerals, students should be able to provide answers for the questions,

What comes next? and *Why?*

It must be kept in mind that students who have come through mathematics classroom settings that fostered *confidence*, encouraged *risk taking* and *use of imagination* are likely to have many correct answers to the simple question, *What comes next?*

For example, one student extended the sequence **1**, **3**, **5**, with the numerals **8**, **11**, and **14**. When asked what he would write next, he responded with **18** and gave an explanation for what he was thinking. It is somewhat sobering to think that had this item been on a test that had identified **7** as the only correct answer it would have been wrongly suggested to this student that he lacks knowledge of the ability to count rationally. This scenario illustrates the need for specific and detailed instructions if it is the intent to elicit one specific response that is identified as being correct.

Visualizing

The *visualization* of numbers is an important component of number sense because it is a pre-requisite for various cognitive aspects of mathematics learning. *Visualization* is required for the development of *mental mathematics strategies*. Tasks that can yield insight into students' ability to *visualize* number can include:

- Briefly showing several fingers on one hand, telling a student to think about all of the fingers on one hand and asking,
  How many did you see? and, *How many do you think you did not see?*

- Briefly showing several fingers on one hand or with both hands, telling the student to think about the ten fingers on both hands and asking,
  How many did you see? and, *How many do you think you did not see?*

- Presenting a name for a two-digit number, i.e., **64**, and asking a student to think of the least number of children it would take to show this number with fingers and explain how this would be done.

- Presenting a name for a two- or a three-digit number, i.e., **59** or **731**, and asking a student to think of the least number of base-ten blocks it takes to show the number and then to name the blocks.

- Using base ten blocks to represent the number for a two- or three-digit numeral, i.e., **37** or **478**, and covering part of the blocks. For example, for **37**, **one ten** and **two ones** are left uncovered. The student is told that blocks were used to show **37** and is asked to describe the blocks that the student thinks are hidden behind the piece of paper. Since different answers are possible for the request, a student may have to be reminded that the number was built with the least number of blocks. A similar procedure is employed for three-digit numerals.

- When base ten blocks are used to show numbers that are products of the multiplication tables, i.e., **12**, **18**, **24**, **32** the student is requested to try and name the double for these numbers.

- When base ten blocks are used to show numbers that are products of the multiplication tables, i.e., **24**, **36**, **48**, **56**, **72** the student is requested to try and name one-half of these numbers?

- A student is requested to explain the different meanings of the digits in numerals that are the same, i.e., **444**.

Flexible Thinking about Numbers

Flexible thinking about numbers is a pre-requisite for developing *mental mathematics strategies* that allow students to teach themselves the basic facts. Indicators of the ability to *think flexibly* include:

- Showing and naming, without referring to addition, the numbers for **four** to **eight** in at least two different ways with fingers.

- Stating several different names for **nine** or explaining different ways of placing **nine** cookies onto two plates.

- Considering only tens and ones and stating at least two different ways of showing and naming the numbers for several two-digit numerals between **twenty** and **ninety-nine**.

- Explaining and naming the results of changing **one hundred** to **tens** or changing **one hundred** to **tens** and **one ten** to **ones** for three-digit numerals.

Explaining the meaning of the digits in a numeral by writing it in expanded form, i.e., **792** or **700 + 90 + 2** or, indicating the meaning of each digit by appropriately labelling it, i.e., **683** or **6hundreds 8tens 3ones** are not indicators of flexible thinking about numbers.

Relating

Do students use the correct language when they make comparisons of numbers or numerals and when they put numbers or numerals in order?

- Do the students use the terms *amount* and *number* correctly? Do they use *amount* for descriptions and comparisons of continuous quantity, i.e., *amount* of money or *amount* of paper and *number* when discrete objects are described or compared, i.e., *number* of coins or *number* of sheets of paper?

- When members of discrete sets are compared, do students use the descriptors *more*, *most*, *fewer*, *fewest*, or *just as many*?

- When the numbers of different sets of discrete objects are compared, do the students use *is greater than*, *greatest*, *is less than*, *least*, or *is equal to*?

- Do the students have a generalization about the place value positions for comparing pairs of two-digit or pairs of three-digit numerals?

- Do the students have a generalization about place value positions for putting several two-digit numerals or several three-digit numerals in order?

- When given a three-digit numeral, are students able to state names for numbers that are: one hundred greater/less than; ten greater/less than and one greater/less than, i.e., **653**; **489**; **790**, **899,** without using pencil and paper.

Estimating

Activities that involve *estimating* number contribute to the development of *number sense* which, in turn, contributes to the ability to *estimate*. Since meaningful assessment of students' ability to *estimate* cannot be based on numerical response data, the strategies that students employ have to be examined and considered.

Key ideas related to estimation include:

- Do students have their own definitions for estimating?

- Do students know the difference between estimating and guessing?

- What *referents* do students use as they explain their strategies for estimating number? For example:
  56 blocks are displayed.
  The request is made to provide an answer for, *About how many?*
  The strategy for arriving at the estimate is to be explained.

- Explaining the estimation strategies that are used for examples like:
  - Estimating the number of words on a page.
  - Estimating the number of pages in a book.
  - Estimating the number of marbles in a jar.

Connecting

Part of making sense of numbers includes the ability for students to *connect* as many numerals as possible to groups of discrete objects that are encountered in their experiences.

- What do students think of when they see numerals or hear number names for some multiples of ten? i.e., **20**, **40**, **50**, **80**.

- What do students think of when they see numerals or hear number names for multiples of one hundred? i.e., **100**, **200**, **300**, **500**.

Calculating Mentally

The ability to *visualize* numbers and to *think flexibly* about numbers enables students to get unstuck when an answer is not known or to reinvent something that may have been forgotten. *Visualization* and *thinking flexibly* about numbers enables students to teach themselves the *basic addition facts* and *basic subtraction facts* and to develop *personal strategies* for addition and subtraction beyond the basic facts.

- The request is made to pretend that an answer for a basic fact is forgotten and to try to think of at least two different ways, other than counting, to calculate the answer, i.e., **8 + 6 = □**; **9 + 7 = □**.
  The request could be in the form of having to explain how calculating an answer could be taught to someone who had forgotten it, with the reminder that ways other than counting should be used. This request may require redirection as indicated by the responses to this type of a request from three different students:
  'Tell them what the sign means. Help them a lot.
  Tell them to use a piece of paper.'
  'I would teach them with a calculator.'
  'I would tell them to try harder.'
  One student's response seems to indicate an awareness of the existence of individual differences, *'Some I will tell the answers to.'*

- The students are presented with an addition equation, i.e., **48 + 39 = □** or a subtraction equation, i.e., **72 – 25 = □**. The request is made to explain how the answers can be calculated without the use of pencil and paper. If the students are successful, they are asked to try to think of another way to figure out the answer in their heads.

Recognizing

In the early grades students participated in activities that fostered the ability to *subetize* or to recognize numbers less than ten without having to count. This ability is not possible for numbers greater than ten. However, students can learn to recognize and name some representations of teen numbers without having to count. For example, when students look at a group of ten fingers, or the fingers shown by one student, and four fingers held up by another student they should be able to report, '*One ten and four or fourteen*' without having to count.

Many experiences with base ten blocks should enhance the ability to report an initial response when a number is briefly shown. This response is independent of any counting and of applying any estimations strategies; it is simply a reaction that can be in the form of:

- *When I looked at the number it made me think of _ .*
- *To me it looked like it was about _.*
- *It looked like less than _ to me.*
- *It looked like greater than _ to me.*

Base ten blocks could be placed on an overhead projector or on a table and then covered with a piece of cardboard. After the students had a brief glimpse at the number they are asked to report or record their initial reaction to,

What number were you thinking of?

Why do you think you were reminded of that number?

These pre-requisite ideas and skills related to the development of *number sense* facilitate the learning of new skills, procedures and the basic facts. Without theses pre-requisites, learning about mathematics can become cumbersome or overwhelming.

Number Sense: To Ten Thousand

The entries for the different components of *number sense* from the previous grades provide suggestions for possible diagnostic tasks. Some of the ideas and activities that are described for the different aspects of *number sense* can also become part of introductory teaching settings as the numeration system is extended to ten thousand – **10 000**.

The key components and properties of the base ten or the *Hindu Arabic Numeration System* include:

- A positional or *place value system* where a digit takes on the value determined by the place it occupies in a numeral.

- The *multiplicative* and *additive principles* mean that numbers are the sum of the products of each digit and its place value in a numeral,
  i.e., **432** = **(4 x 100)** + **(3 x 10)** + **(2 x 1)** or **400** + **30** + **2** or **432**.

- *Zero* as a name for the number zero and as a *place holder numeral.*

- A *grouping pattern by 'threes'* for reading, *ones – tens – hundreds*, for naming and reading numerals. Groups of three digits are separated by spaces rather than commas since these are used as a decimal point in some places.

These components and properties make it easy to learn how to read, memorize and write literally millions of numerals without too much effort.

Types of Activities and Problems

The main intent is to focus on selected types of activities, problems and settings that are deemed conducive to the development of *number sense* and *numerical power*. The key role played by a teacher is reiterated.

Visualizing

It is advantageous to use manipulative materials that keep contributing to the fostering of *visualization* of number.

<u>Base ten blocks</u>: These blocks clearly indicate the relationships of the base ten numeration system. They are essential for activities designed to contribute to fostering the *visualization* of numbers, at least to ten thousand.

Other devices like a pocket chart, an abacus or a place value chart are abstract and do not contribute to *visualization* of number. These devices can be thought of as 'counters' of the numbers that students are to *visualize*.

Some references make use of different colours for different place value positions in pocket charts or on an abacus. This representation can be detrimental since number is independent of colour. When colour is used it can distract from the intended outcomes. For example, some students may be tempted to make comments that make reference to the colours rather than to the numbers, i.e., '*Ten greens are changed for a red.*'

An attempt to foster *visualization* that involves placing pieces of different sizes into different place value positions of pocket charts is mathematically incorrect. Once place value charts have been labelled with headings:

Thousands – Hundreds – Tens – Ones

they become or are 'counters' and it is mathematically incorrect to place tallies or blocks other than ones, as counters of ones, tens, hundreds, etc., into the different parts of such a chart.

- **Activities with base ten blocks – Sample tasks and settings**
  The blocks or pictures of the blocks are used to show numbers and the students are asked to record the numerals for each representation. A request to rewrite numerals in expanded form can contribute to fostering *visualization*. To point out the role of zero as a place holder and why zeros are needed, examples like, 'two one-thousand blocks and four ones' are shown.

  Students who have never used base ten blocks come up with some intriguing responses when shown a one thousand block and are asked, *How many?* One very observant student responded with, '*six hundred.*' He was correct, since he was knocking on a hollow plastic model with six faces of one hundred each.

  When a stack of ten hundreds is exchanged for a one thousand block, some students will be able to keep in mind the subdivisions of the hundred blocks and visualize how many ones there are in one thousand. To foster *visualization* a reminder to think about each layer as one hundred may be required.

  After a three-digit or a four-digit numeral is presented i.e., **479**; **5 281**, types of requests can include:
  - State and name the least number of base ten blocks it would take to show the number.

- State and name the greatest number of base ten blocks it would take to show the number.
- Show several other ways of representing the number with base ten blocks and to explain the answers.

A three-digit or a four-digit numeral is shown with base ten blocks, i.e., **372**; **4 957**. The numeral and only part of the blocks are shown.
For example:
- For **372**, **1** hundred and **1** ten are shown.
- For **4 957**, **1** thousand, **1** ten and **2** ones are shown.

The following requests are made:

What blocks are hidden if each number is shown with the least number of blocks? Explain your answer.

- **Print-outs to foster visualization**
  Computer generated diagrams of stars showing **10**, **100**, **1 000**, **10 000** dots can be displayed on a bulletin board or wall.
  These print-outs can be referred to as number names are discussed.
  For example:
  - Whenever a one thousand block is used, the print out will be a reminder of how many ones there are in the block.
  - Pointing to the one thousand print-out as number names in the thousands are discussed can contribute to *visualization*.
  - When numerals in the ten thousands are used, the appropriate print-out can be pointed to.

  An attempt to illustrate and *visualize* the magnitude of the numbers one thousand and beyond can include sharing the answers to problems of the following type:
  - Assume you are able to say one number name every second, about how many minutes or hours would it take to count to one thousand and to ten thousand?
  - About how many pages in a book of your choice would it take to read one thousand words and ten thousand words?
  - About how high would a stack of one thousand and ten thousand pennies or one thousand and ten thousand mathematics books be?
  - About how many times would one thousand and ten thousand steps take you around the outside of the gym, around the school or around a soccer field?

Flexible Thinking about Numbers

Thinking flexibly about numbers implies knowing that a number can be shown and named in different ways. This knowledge is a requisite for developing *personal strategies*. *Thinking flexibly* about numbers enables students to rename numbers to suit their preferred way or ways of using them as pencil and paper or mental calculations are carried out. The renaming of numbers while performing calculations is done without the use of the non-mathematical terms borrowing and carrying.

- A number made up of more than ten tens or more than ten hundreds is displayed. The request is made to do the necessary trading to show the number with the least number of blocks and then record the standard numeral for the number.

- Play money (**$100**s; **$10**s; and **$1**s) – actual or sketches – is used to show an amount of money, i.e., **$341**. The request is made to use the given denominations and to think of and sketch this amount in at least four or more different ways.

- For a given amount of money in the thousands the request is made to explain at least two different ways of showing the same amount.

- For a given amount of money, i.e., **$5649**, one of the one thousand dollar bills is to be exchanged for different denominations. The request is made to record some of the possible equivalent amounts of money.

Relating

The language used to identify, compare and order numbers, representations of numbers, and numerals includes such term and expressions like: *more, fewer, as many, same number of, greater than, less than, close to, far apart, about the same, odd, even, square, triangular, ones place, tens place, etc., digits, sum of digits, difference of digits.* As new terms like: *multiple, factor, divisible, prime, composite* become part of mathematics learning, these are added to the list to be used for writing tasks and for discussions.[5] [6] [7]

- **Writing and Talking about Numbers and Numerals**
  Writing tasks provide opportunities for thinking, thinking about thinking, editing, decision making, evaluating, comparing alternatives, aspects of problem solving and as the example mailed in by one student illustrates, use of imagination. As creations are shared with classmates mathematical language is used and developed.

  Before the opportunity to write about a mystery number or numeral is given, several examples should be examined.
  For example:
  As the numerals from **840** to **879** printed in four rows on the chalkboard are examined, the following hints about one of these numerals are presented, one at a time. After each hint the request is made to suggest several number names that could be the mystery numeral and then name those that are ruled out. The latter are crossed out or covered up.

  Mystery Numeral
  The numeral does not have a 0 or an 8 in the ones place.
  The numeral is not between 841 and 848 and not between 871 and 878.
  The numeral does not have a 7 in the tens place.
  The numeral is a name for an even number.
  The sum of the digits in the numeral is less than 18.
  The sum of the digits is greater than 15.
  One of the digits is a name for an odd number.

  - **Accommodating Responses – Providing Assistance**

    In response to the request to write hints about a number or numeral, there will be students who will write just two statements which will make it possible to identify a mystery numeral from a list of numerals. To avoid this occurrence, the request can be made to write at least four hints and to use different terms or phrases for each of the hints.

Observations made in classrooms show that the majority of students enjoy writing about mystery numerals. They enjoy sharing what they have written, especially when the hints they had written were projected onto a screen for everyone to read. Many appeared proud when they had others deal with the information they had created. Increased specificity and improvement in writing was noticed when students were given several opportunities to write about a mystery numeral over a three week period.

During visits to different classrooms the request has been made to try and think of more riddles and to forward these. The following example that was received in the mail illustrates the value of this type of an activity. One boy in grade four included newly learned vocabulary as part of what he called a *Math Riddle* about a number name on the **99**-Chart:

*The number – below **100***
*– above **50***
*It's a multapul (sic) of **2** and divisabale (sic) by **3***
What is the number?
answer on back
picture on back for you

Along with the answer **66** on the back appears a multapul (sic) creature, a combination of a horse and a goose – called a *Gosse*.

If classrooms take advantage of parent help, some direction is required. Some parents tend to be too helpful by correcting students or by phrasing sentences or by writing sentences for them as attempts are made to write about a mystery number. This can deprive students of a valuable learning experience when they respond to questions and comments from their classmates while statements about a mystery number are shared. The use of appropriate and correct mathematical language is a gradual process and making statements for students or providing them with terms or phrases does not shortcut this process.

- **Special Challenges and Requests**

A mystery number or mystery numeral setting can easily be created by having students generate their own list of fifteen or twenty four-digit numerals and write about one of the members of the list.

As the hints that are provided by different students are examined, questions of the following type could be posed:[6]

- *Is finding a mystery number different when the last hint is read first or when the hints are read in reverse order?*
- *Do you think one hint gives more information about the mystery number than the others? If so, why is that the case?*
- *Do you think one hint gives less information about the mystery number than the others? If so, why is that the case?*

Writing about a mystery number can include requests for special types of hints. The readers or listeners are challenged to identify these hints.
For example, requests for special types can include:

- *Hints about the number that contain extra or irrelevant information.*
- *Hints about the number that do not help to identify it.*
- *Hints that differ from previous hints, but provide the same information.*

Writing hints about a mystery number or a mystery numeral that are *not true* requires a way of thinking that can be a new experience. Prior to writing hints of this type, and initial attempts could be tried with a partner, examples should be examined. As is the case for hints that *are true*, one hint is presented at a time and each time the questions are:

What could the number be?

What number or numbers can be ruled out?

For example, twenty rectangular regions or doors labelled **0** to **19** are sketched on the chalkboard. Each hint for finding the secret door or entrance is not true:

The numeral on the door names:

An even number.

A number is greater than fifteen.

A multiple of three.

A number less than the number of fingers on both hands.

A number greater than one dozen.

Confidence and a feeling of importance can be fostered when transparencies of the hints are prepared and the authors show these hints, one at a time, and orchestrate the discussion of the group. It is during these discussions that the questions that are raised and the comments that are made by the participants can result in suggestions for changes or revisions and thus provide opportunities for the authors to think about thinking.

- ***Naming Numbers***

  A numeral is presented, i.e., **462**; **7839**. The requests that are made can include:

  State or record the numerals for numbers that are:

  One more; one less,

  Ten more; ten less,

  One hundred more; one hundred less, and

  One thousand more; one thousand less.

  Some of the numerals that are presented should have zeros and nines in the different place value positions.

- ***About 'About'***

  The students are asked to look at a numeral, i.e., **7643**, use the word *'about'* to identify which member of a pair of numerals the numeral is closer to and explain their thinking:

  About **10000** or **1000**;

  About **8000** or **7000**;

  About **7700** or **7600**;

  About **7650** or **7640**.

  A partial number line can be used. Each request includes placing a given numeral on the line and explaining the thinking for the placement.

Estimating

Whenever possible opportunities need to be presented to continue to develop and use estimation strategies. Reminders may be required about the difference between guessing and estimating. Estimating number means that the number for a group of objects or things is known and is used as a *referent* to come up with an answer to the questions:

Do you think there are more than or fewer than ... (the referent)*?*

About how many are there? What do you think?

The responses for the last two questions should be to the nearest ten or hundred depending on the *referent* that is used. The print outs that were described as part of fostering *visualization* can serve as *referents* for some estimation tasks.

- A pre-counted number of blocks or chips, i.e., **54**, is shown with an overhead projector (or as dots on a piece of cardboard). The request is made to use the fingers on both hands, or ten, to come up with an answer for:

  About how many tens do you think are shown? or,

  Into how many groups of ten do you think these counters could be grouped?

  The responses are to be recorded as: *About* __ *tens.*

  To reinforce the difference in meaning between estimating and guessing, each task could include the recording of a *lucky guess* __ for the actual number of counters in a group.

- One hundred names are highlighted on one page of a telephone book. A request is made to record estimates for:

  About how many hundreds of names do you think are on one page?

  About how many hundreds do you think are on two pages?

  and to be ready to explain the strategy that was used.

- A stack of 100 envelopes could be shown to elicit responses for estimates for how many envelopes there are in a large box.

  Tasks of the following type can be used to gain some insight into *sense of number*. A group of pre-counted blocks or dots, i.e., **46**, is shown and three choices are presented along with the request to select one that tells about how many there are.

  Which one would you choose and why that one?

  About 20 About 50 About 90

  Rather than three choices, two could be presented. As experience is gained with numbers, estimating numbers and *number sense* develops, the range of the choices that are presented is reduced.

- A collection of thumbtacks or paperclips is used to show the number **260**. An initial request can be an invitation to select a choice for:

  About how many thumbtacks? What do you think?

  About 10 About 100 About 1 000.

  This task may seem somewhat trivial for an adult, but if two of the choices are selected, they provide a lot of information about *sense of number* and provide valuable information for intervention.

A second part of this setting can be similar to the task with the pre-counted blocks. Three choices are presented along with the request to explain the strategy used to make the decision:

About 100 About 300 About 600

Meaningful assessment of students' ability to estimate cannot be based on numerical responses that students report or have recorded. The strategies that were employed to arrive at estimates have to be examined and considered.

Connecting

Connecting multiples of one thousand to ten thousand to experiences outside the classroom requires an awareness of events in the community and being on the look-out for reports in the newspaper that are meaningful for students.

- A choice format can be used to provide the opportunity for reflection about numbers as attempts are made to *visualize* which of several choices is meaningful and which choices are not.
  For example:

  About how many people were in the gymnasium for the concert?
  30 300 3 000

  About how many people were in the movie theatre?
  20 200 2 000

  About how many people showed up at the Terry Fox Run?
  10 100 1 000

  Other questions along with appropriate choices and distracters can include:
  About how many people live in our town?
  About how many people were at the game?

- Requests to construct personal lists can begin with the tasks of listing:
  - The multiples of one hundred to one thousand.
  - The multiples of one thousand to ten thousand.

  The challenge is presented to begin to collect entries of descriptions of groups of people, animals or objects that make sense in terms of the students' experiences. At some point, time is set aside to compare the lists. A display of the lists could be prepared.

Number Sense: Beyond Ten Thousand

It is a challenge to try and help students make sense of numbers that are difficult, if not impossible to *visualize*. One goal about these numbers is related to the realization that one hundred thousand and one million are large numbers and it may be wise to think for a while before making reference to numbers like these somewhat carelessly during conversations.

Types of Activities and Problems

- One possible way to attempt to impress upon students how much one hundred thousand and one million are is by extending the print outs that were referred to when numbers up to ten thousand were examined. A display of computer generated dots in the hallway of ten, hundred, one thousand, ten thousand, one hundred thousand, and one million can serve as a visual reminder every time these numbers are looked at.

- Attempts to connect one hundred thousand and one million by trying to solve problems about objects, actions or events from the students' experiences can result in a conclusion like, *Wow, these numbers are' humongous'!*
  For example:
  - *How many birthdays would you have in one hundred thousand days and in one million days?*
  - *About how thick might a book be if it had one million pages?*
  - *About how many pages would it take for printing your name one million times?*
  - *About how long would it take to count to one million or count one million one dollar coins?*
  - *About how long would it take for the heart to beat one million times?*
  - *About how long would a line up of one million people, bicycles or cars be?*

  If students work in pairs and try to find the answer to one or two questions that are of interest to them, they should be asked to share the thinking and the strategies that were used with their classmates.

- Requests are made to write hints about a mystery number in the millions and share these with classmates. A challenge can be introduced by suggesting that all of the hints about a mystery number are to be false.

- A picture of a stadium, or a hockey arena, filled to capacity (or partially filled) with fans is displayed. The request is made to come up with estimation strategies for the number of fans in attendance. The strategies and results are compared.

- After the multiples of one hundred thousand are listed, the request is made to identify entries beside each multiple that tell who uses numerals or numbers like these and when. A reminder that the word *about* is appropriate for the entries in the list may be required.

- A list of six to ten numerals in the hundred thousands is presented. The request is made to think of the numbers for these numerals in order from least to greatest and to make up a rule for ordering any set of numerals for someone who does not know how to do this. The rules are compared.

- The request is made to prepare a written response for someone who claims that number names with nines always name the greatest numbers. The responses are shared and compared.

Number Sense: Fractions, Decimals, Percent and Integers

The main goal is to consider key components of fraction –, decimal –, percent – and integer *number sense* and to illustrate each of these components with a few sample tasks. These components of *number sense* are essential for future learning since they will enable students to:

- Develop personal strategies for computational procedures.
- Derive their own rules or generalizations for performing computations.
- Reinvent rules for carrying out calculations that may have been forgotten.

Fraction Number Sense

The key components of *number sense* with whole numbers give an indication of what is involved in fostering *number sense* with fractions. Recognition and naming, or subetizing, is not part of fractions. However, students should be able to tell, at a glance, whether or not regions are divided into equal parts, which at the introductory stage means congruent parts. This definition of fraction can create a dilemma because outside the mathematics classroom the term is used by many people to refer to pieces or parts of regions that are not equal or not congruent.

When sets of discrete objects are used for the teaching about fractions, students have to learn to consider the number of objects rather than the size and/or shape of the objects. Fractions are used to name points and distances on the number line.

Visualizing

Activities that contribute to the *visualization* of fractions include: counting pieces, shading, cutting, folding, sketching, building and naming. After the convention of **a** over **b**, and the meaning of each part are introduced, reading every fraction in two ways, i.e.,

4/5 – *four out of five equal pieces or parts* and *four fifths*

contributes to fostering *visualization* and will enable students to represent any fraction concretely and/or pictorially. Students should be able to state that the 'equal' that is part of the reading refers to same size and shape, or congruency.

- Activities that involve cutting, folding and sketching can lead to the conclusion that one-half of the same region can be shown in different ways. Each of these ways of showing one-half looks different but there is something the same about them. Students need to conclude that if someone talks about one-half, the size of the piece that is described is known but the shape may not be. At this stage of learning, students will also realize that one-half can mean different things, depending on the size of the region that is considered.

- *Visualization* is fostered when parts of shaded areas of regions or some arrangements of blocks are examined that are designated to represent certain fractions. The challenge of trying to *visualize* and identify the whole is presented.

For example:

- *Different arrangements of three adjacent shaded square shaped regions are identified as three-fifths of a whole. What could the whole look like? Prepare sketches. Explain your thinking.*

- *Different arrangements of four adjacent blocks of the same type and size are labelled as being four-tenths of a whole. What could the whole look like? Prepare sketches. Explain your thinking.*

Flexible Thinking about Fractions

One way fractions differ from whole numbers is that many more names can be generated for every fraction; in fact students will discover that each fraction can have as many names as there are counting numbers or an infinite number of names. The other exciting or amazing difference is that for every pair of fractions, another fraction that is between the two can be identified and named.

- On several copies of rectangular regions of the same size and shape, students are asked to show by folding and shading:

  $\frac{1}{2}$, $\frac{2}{4}$, $\frac{3}{6}$ and $\frac{4}{8}$.

  What is the same for each shaded region?
  What is different for each shaded region?

- After recording several different names for one-half, the request is made to try and make up and record a rule.
  What would your rule be for writing other names for one-half?
  The rules are shared and compared.
  How many different names do you think can be written for one-half? Explain your answer.

- After at least one other name for several fractions is recorded, the request is made to make up and record a rule.
  What would your rule be for writing other names for any fraction?

Relating and Estimating

A key part of *relating* and *estimating* involves the use of *benchmarks* on a number line. After the use of *close to* **0** or *close to* **1** *whole*, $\frac{1}{2}$ is introduced. As experience is gained, the benchmarks $\frac{1}{4}$ and $\frac{3}{4}$ can be introduced. The use of the phrase *out of* as part of reading fractions fosters *visualization* and contributes to the ability of placing fractions on a number line.

- A list of fractions less than one whole is presented. For example:

  $\frac{3}{4}$, $\frac{1}{3}$, $\frac{1}{4}$, $\frac{7}{8}$, $\frac{2}{3}$, $\frac{3}{8}$.

  After several fractions from the list are identified on a number line, the request is made to write a statement about,
  How can you tell whether a fraction is closer to **0** *or closer to* **1** *by looking at the fraction?*
  The statements are shared and compared.
  The activity and a revised request can be repeated for a new list of fractions and three benchmarks: **0**, $\frac{1}{2}$, and **1.**

- The request is made to order several like fractions from least to greatest and to record a rule for ordering fractions with the same denominator.
  For example: $\frac{8}{10}, \frac{4}{10}, \frac{7}{10}, \frac{5}{10}, \frac{2}{10}, \frac{6}{10}.$

  The rules are shared and compared.

- Six or more different unit fractions are presented.
  For example: $\frac{1}{10}, \frac{1}{5}, \frac{1}{2}, \frac{1}{6}, \frac{1}{100}, \frac{1}{3}.$

  After the fractions are ordered from least to greatest a rule is to be recorded for ordering fractions with a numerator of one. The rules are shared and compared.

  Do you think your rule would work for any group of fractions
  that have the same numerator?
  Explain your answer.

- Several unlike fractions are presented.
  For example: $\frac{3}{4}, \frac{99}{100}, \frac{4}{10}, \frac{1}{2}, \frac{1}{5}, \frac{5}{8}.$

  The request is made to try to order these fractions from least to greatest. As the results are presented, the strategies used to make the decisions are discussed and compared.

The types of activities above can be repeated with improper fractions.

Connecting

The questions, *Who uses fractions? When? Where? Why?* need to be part of ongoing activities with fractions. For future transfer and connecting to decimals, activities with the denominators **10**, **100** and **1 000** should be part of the activities.

To accommodate students who may lack fine auditory discrimination, a chart showing the following information needs to be prepared and displayed:

tens - tenth – tenths
hundreds - hundredth – hundredths
thousands - thousandth – thousandths

The appropriate terms are pointed to during discussions or students are asked to point to the appropriate part of the chart as they share information.

Decimal Number Sense

Some mathematics programs introduce students to decimals as an extension of the Hindu Arabic Numeration System, while others base the introduction on an intuitive knowledge about fractions with denominators **10**, **100** and **1 000.**

Visualizing

Visualization of decimals can be fostered by covering the faces of the one thousand block and the one hundred flat of the base ten blocks with paper or light cardboard. These covers hide the subdivisions that can be distracting at the introductory stage, but are also meaningless or undefined as far as the understanding of decimals at the introductory stage is concerned.

The base ten blocks can become a concrete representation that aids *visualization* of decimals to thousandths. Student are asked to pretend that the smallest block, or a one, is placed under a magnifying glass so strong that it increases in size to become as big as the one thousand block whose faces have been covered. This block is now labelled and used as **one whole**.

Tenths: To introduce tenths and to foster visualization nine horizontal segments are drawn on the one whole to show ten equal parts or ten tenths.

Hundredths: When hundredths are introduced nine segments are drawn on the covered face of the flat or the one hundred block.

Thousandths: To foster visualization of thousandths, the paper or cardboard cover on the one whole is removed to show the thousandths.

Convention: The convention used for recording and reading decimals also contributes to fostering *visualization*.
For students who may lack fine perceptual discrimination, the recording of a zero prior to the decimal point for decimals less than one serves as an advance organizer for the decimal point. Reading the decimal point as *and* followed by fraction names for three place value positions to the right of the decimal point *connects* to previous knowledge about fractions and fosters *visualization*.

For students who may lack fine auditory discrimination a chart that can be pointed to is required. As is the case for fractions, this chart should show:

tens - tenth – tenths;
hundreds - hundredth – hundredths; and
thousands - thousandth – thousandths.

During discussions appropriate parts of the chart are pointed to.

Flexible Thinking

The base ten blocks can be used to illustrate the effects of annexing a zero to a decimal. The answers to the following questions can be illustrated by showing each decimal with the appropriate base ten blocks:

*How are **0.2** and **0.20** and **0.200** the same?*
How are they different?

The importance of proper and correct language can be illustrated and reinforced by requesting students to select a decimal i.e., **0.2**, and then to illustrate the difference between 'adding zero' and 'adding on zero' to the decimal.

The grouping by ten property of the Hindu Arabic Numeration System can be illustrated and reinforced with the base ten blocks.
For example, students are asked to show and record the standard names for examples like:

- Sixteen tenths.
- Twelve hundredths and six tenths.
- Fifteen hundredths and eleven tenths.

Relating and Estimating

The language for relating decimals is the same as for whole numbers and numerals. Many of the activities suggested for whole numbers as well as for fractions can be adapted for decimals.

- A list of decimals is presented. After the placement of several decimals less than one has been identified, the request is made to write a statement about how it is possible to tell whether a decimal is closer to **0.0** or closer to **1.0** by looking at a decimal.

- Three *benchmarks*: **0.0** - **0.5** - **1.0** are considered for placing decimals on the number line. Visual reinforcement for the distances on the number line can be created by placing the appropriate number of base ten blocks next to the points showing **0.5** and **1.0**.

- The placement on the number line with three *benchmarks* is repeated with decimals greater than **1.0**.

- After ordering several decimals from least to greatest the request is made to write a rule for completing such tasks. How is it possible to tell that the order is correct? The rules are shared and compared.

- Riddles about a *Secret Amount of Money*, a *Secret Decimal,* or a *Mystery Measurement* can be solved and created. The questions and problems posed for these tasks can be the same as those posed for writing about numbers and numerals.
  For example:

  Number of Coins Mystery
  Someone has **50 cents** and has **12 coins**.
  There are dimes, nickels and pennies.
  There are just as many nickels as there are pennies.
  What are the coins?

  The solution strategies that were used are shared and compared.

  Opportunities to discuss strategies and to use the language that is part of *relating* decimals are provided if initial writing tasks that involve creating a mystery are undertaken with a partner.

  Solving and creating riddles of this type can contribute to fostering *flexible thinking* about number.

  Riddles about amounts of money are similar to the examples described for whole numbers. These types of riddles are appropriate after decimals to hundredths are introduced and a decimal point is used to record amounts of money.

Amount of Money Mystery

The amount of money is between one dollar and two dollars.
The sum of the three digits in the amount is equal to twelve.
The name for the amount is not a name for an even number.
The name for the amount is not a multiple of five.
The number of pennies in the amount is greater than five.
The amount is close to one dollar and fifty cents.
What is the amount of money?

Name something that you think could be purchased for this amount of money.
Did anyone identify the same object or thing?

For riddles about decimals or measurements, several decimals could be listed and hints are provided about one of the entries in the list.

Measurement Mystery

1.00 m 1.08 m 4.62 m 8.80 m 9.01 m 9.19 m 18.59 m 23.17 m

There is not a one in the tenths place.
There is not a zero in the hundredths place.
The measurement is greater than nine metres.
The sum of the digits is less than twelve.
The secret measurement is?

Name something that you think is about this tall or this long.
Did anyone identify the same object?

Accommodating Responses:
The first time students write their own riddles it is likely that some will follow the same format of the last example that was attempted. These students can benefit from working with a partner and the challenge of trying to make use of terminology or phrases that have been listed for them on a chart.

Connecting

The following questions need to be part of ongoing activity settings:

Who uses decimals? When? Where? Why?

Students should know numerous examples of things that are thought of as being partitioned into ten, one hundred and one thousand equal pieces or parts. Some of these are not easy to *visualize* as the divisions that are part of a metre.
For example:

- Ten equal parts of a cent as is shown at the gas pumps. What is measured to the nearest hundredth and nearest thousandth at these pumps?
- A second divided into ten, one hundred and even one thousand equal parts for timing of sport events.

Percent Number Sense

Before students look at percent, they should know that a numeral in the form **a** over **b** ($\frac{a}{b}$, where **b** is not equal to zero) requires context before it can be read and interpreted. It could be a fraction, it could refer to division, or it could be a comparison ratio or a rate ratio. Knowledge of ratio is a requisite for the introduction of percent

Connecting to ongoing and previous learning will allow students to know how to read $\frac{3}{100}$ as *three per hundred* or *three per-cent* and then write **3%** and **0.03**.

Visualizing

The flat block, or the one hundred base ten block is well suited for fostering *visualization*. Different parts of the flat block, or square shaped regions equal in area to the flat block, can be shaded or portioned off to have students label these parts as fractions, ratios, decimals and percent. Recording a given percent as a decimal and shading an appropriate region is one indicator of ability to *visualize*.

Relating and Estimating

- As shaded parts of square shaped regions equal in area to the one hundred base ten block are presented, students are asked to use the following *bench-marks* to estimate a shaded area that is shown and then reporting the answer as percent:
  - More than **50%** ($\frac{50}{100}$ or **0.50**); less than **50%**.
  - About **25%** ($\frac{25}{100}$ **or 0.25 or** $\frac{1}{4}$), about **50%**, and about **75%** ($\frac{75}{100}$ or **0.75** or $\frac{3}{4}$).
  - About **35%** (about $\frac{1}{3}$) and about **65%** (about $\frac{2}{3}$).

- Shading is used in sketches to show different amounts of liquid in soft drink cans. The request is made to record estimates as percents for the shaded part of each container.

  Explain the strategy that was used to arrive at the estimate.
  What makes it easy to record an estimate for each part that is not shaded?

- Sketches of rectangular regions are presented. The request is made to write sentences that explain the strategies used for shading parts of the regions which show:

  About **10%**. About **90%**. About **30%**. About **80%**.

- A request is made to prepare a list of percents that are easy to estimate and to write sentences that explain why that is the case. The sentences are compared.

- Someone reported that a can of juice has about **40** gulps or mouths full to drink. The request is made to explain how sketches could be used to calculate the answer for:

  About how many gulps in **25%** of the can?
  What percent of the can is equal to about **30** gulps?

  If someone reported that **10** gulps or mouths full is **20%** of a juice can, what sketch could be drawn to show how many gulps there are in the whole can for this person?

Connecting

Groups of two or more students are requested to collect and list information for the question:

Where have you seen percent used and listed?

After the lists are shared and compared, the challenge of trying to illustrate each member of the list with a sketch is presented. The sketches are compared and labelled samples are displayed.

Many times coaches and athletes can be heard to make statements like,

To win we have to give one hundred ten percent.

We won because we gave it more than one hundred percent.

Explain your reaction to statements like these. What questions would you ask of these athletes and coaches?

Integer Number Sense

Visualizing and Connecting

Integers are an extension of the system for whole numbers. Models for integers are less concrete than those for fractions and decimals. However, reference can be made to many different everyday events or situations dealing with magnitude as well as direction that make use of negative integers.

Statements about: loss and profit; temperature; golf scores, football and hockey statistics; sea level; and electronic charges can all involve integers. Calculators also show negative integers.

A thermometer clearly illustrates how opposites – positive five and negative five – are equal distance from zero as well as how a negative number can be represented by an arrow and direction. For example, starting at zero on a number line, an arrow of five intervals to the right shows positive five and an arrow of five intervals to the left shows negative five.

Different colours, red and black, can be used to illustrate electrical charges as well as the difference between loss and profit.

Relating

The number line illustrates how order is established for integers. Since numbers increase as moves are made from left to right, the same order relations hold true for integers.
Thus, the number line illustrates that any negative integer is less than zero.

The number line illustrates that moving to the right shows that negative four (**-4**) is less than negative three (**-3**) and greater than negative five (**-5**).

Listing several negative integers, identifying one as the *Mystery Integer*, and writing hints about it provides an opportunity to reinforce the notion of order illustrated on the number line.

Flexible Thinking about Integers

The notion of 'net', i.e., net worth or net score, needs to be introduced in order to lead students to realize that any integer can, as was the case for whole numbers, fractions and decimals, be represented in different ways. That means that a net worth of $3; a net change of negative two; or a hockey game statistic of plus two could all be represented in several different ways.
For example:

- A net worth of **$3** could mean having **$4** and owing **$1**; having **$5** and owing **$2**; or having **$6** and owing **$3**.
- A net worth of negative **$2** could mean having **$5** and owing **$7** or having **$6** and owing **$8**.

This *flexible thinking* about integers will assist students as they attempt to derive their own rules for the addition and subtraction of integers.

Number Sense: The Key to Numeracy and Numerical Power

The chapter includes many ideas and examples that illustrate why *number sense* is the key foundation for *numeracy*. *Number sense* is essential for developing the ability to *reason mathematically*. *Number sense* enables students to develop *personal strategies* and coin *personal generalizations*.

The following scenarios illustrate the contribution *number sense* can make to *numerical power*.

- The ability to solve an equation like **7 + 5 = □** in different ways is indicative of the *ability to visualize* numbers and to *think flexibly* about numbers; an ability that transfers to other calculation procedures.

  Another name for five is three and two.
  I use the three to go to ten and then two more gives the answer **12**.

  There is a gap of two between the five and the seven.
  If I take one from the seven and add it to the five,
  I have a double that I know the answer for;
  six plus six is equal to twelve.

  Another name for seven is two and five.
  The two fives are equal to ten and two more is equal to twelve.

  Number sense enables students to develop *mental mathematics strategies* which enable students to learn the *basic facts*.

- *Number sense* along with a *conceptual understanding* of the operations enables students to develop and employ *personal strategies* for computational procedures beyond the basic facts. *Number sense* eliminates the need for memorizing rules or steps.

Knowledge of the meaning of the digits, the ability to *visualize* the numbers that are named and the ability to *think flexibly* about the numbers enables students to suggest more than one way of adding two three digit numerals like **256 + 472 = ☐**. Students will be able to think:

I can start anywhere as long as I keep track of the digits
that have the same meaning or value.
*For example, there are **6** hundreds, **12** tens and **8** ones.*
*Another name for this is **728**.*

- Operations with fractions are not part of the elementary mathematics curriculum. The intent here is simply to illustrate the power of *number sense* along with *conceptual understanding*. *Visualization* and *flexible thinking* along with *conceptual understanding* of the operations make it possible to simulate the action stated in an equation and to arrive at a generalization about the suggested action.
  For example, students will be able to simulate the additive action for examples like,
  $\frac{3}{10} + \frac{6}{10} = \square$ and $\frac{1}{2} + \frac{1}{4} = \square$.
  After the answers are recorded, the students are requested to make up a rule for adding fractions..

  Subtraction of fractions can be dealt with in a similar way. Thinking of multiplication as *groups of* and division as *taking away groups of* for divisors that are fractions and as *equal sharing* for divisors that are whole numbers will enable students to derive their own computational procedures.

At the end of a lecture a teacher-to-be requested the rule for subtracting negative integers, a topic assigned by the sponsor teacher for a weekly lesson. The equation he recorded on the chalkboard involved the subtraction of negative five from negative two. Responses to questions indicated an inability to reinvent the rule that was forgotten. Integer *number sense,* which includes the ability to think of and name integers in different ways, would have made this reinvention possible. This scenario illustrates one of the greatest disadvantages of learning without *conceptual understanding* and a focus on *number sense*.

Number sense is the key for making mathematics meaningful. Failure to develop *number sense* will mean that many students will end up 'muddling through mathematics without the appropriate attitudes and abilities'[8] and it can happen that, 'mathematics may filter students not only out of careers, but out of school itself.'[9]

Assessment Suggestions

Since *number sense* is an essential component of mathematics learning, assessment data need to be collected and shared with the parents. The collected data have to be translated into specific statements that can be interpreted in the same way by everyone who reads them. The latter can be a challenge. It may be tempting to tell someone that a student has 'good number sense' or that a student 'lacks number sense', but since these statements lack specificity they are meaningless.

As students participate in small group discussions, as they read their hints about mystery numbers and numerals, and as they share and describe their strategies for computational tasks, specific evidence about the components of *number sense* will surface. Whenever possible these indicators of *number sense* need to be recorded. This record can be a combination of phrases and entries on a checklist that become part of a portfolio.

Visualizing

Indicators of ability to *visualize* the numbers for numerals can become evident as:

- Calculation procedures are explained.
- Hints for riddles or mystery numbers or numerals are created, recorded or shared.

Flexible Thinking

Did the explanations for calculation procedures used yield any indicators about being aware that numbers can be represented and named in different ways?

Relating

Was the correct terminology used when comparisons of numbers were made and when statements that are part of riddles or mysteries were created? Did the terminology that was part of the new mathematics that was learned become part of the oral and written language?

Estimating

Was there evidence of the use of *referents* or *benchmarks* as estimation strategies were explained?

Connecting

Did any statements uttered during discussions show any evidence of ability to connect numbers or numerals to events or actions outside the mathematics classroom?

Reporting

Since *number sense* is an important goal of the mathematics curriculum for all grades and it is a requisite for success with learning mathematics, parents need to be informed about this aspect of mathematics learning. A way must be found to inform parents of what *number sense* is and why it is a very important part of mathematics learning. Once this is done, possibly via a newsletter or a presentation it facilitates report writing.

For example, parts of reports could read:

- The data that have been collected include many indicators of the important aspects of *number sense*. These indicators, which are part of the portfolio, suggest that there should not be any difficulties with understanding the mathematics that is being learned and will be learned.

- The data collected and recorded in the portfolio show several indicators of *number sense* with respect to being able to *relate* numbers and *connect* these numbers to everyday experiences. The goals of some of the current activities and problems are to contribute to fostering *visualization* of numbers and *flexible thinking* about numbers.

- At the beginning of the year very few indicators of *number sense* were observed and collected. Recently quite a few indicators were observed, noted and entered into the portfolio. This shows that *number sense* is developing.

- The data collected with respect to *number sense* and entered into the portfolio are indicative of growth with respect to this important part of mathematics learning.

For Reflection

1) What major points would you include in a presentation to parents about the importance of fostering *number sense*?

2) What would you say to someone who states that there is not enough time to allow for a focus on developing *number sense*?

3) How would you respond to someone who asks for several examples of test items that could yield indicators of *number sense* and give some insight into the presence of *number sense*?

4) What suggestions would you make to someone who asks what might be done in the home to foster the development of *number sense*? What ideas would you make about how homework might accommodate reaching this important goal?

5) Why are special skills or why is pedagogical content knowledge required to foster the development of *number sense*? What are some aspects of this knowledge?

6) During a conversation with several teachers who teach in a middle school the observation is shared that many of their students 'lack *number sense*.' What questions would you ask of these teachers? What suggestions would you make? How would you address their concern about possible reasons for this lack of *number sense*?

Chapter 3 – Basic Multiplication and Basic Division Facts

Concerns about Lack of Knowledge of the Basic Multiplication Facts

During meetings with representatives from one Ministry of Education it was reported more than once that the most frequent complaint received from parents dealt with students not knowing their *basic multiplication facts*. Many teachers from the junior and senior high schools have made comments about students lacking this knowledge. During presentations to parents and grand-parents a comment like the following will be shared by more than one person,

'When we went to school we knew our multiplication facts and we knew how to recite these quickly.'

Some grandparents will make reference to the superior knowledge of children in other countries.

Possible Issues Related to the Curriculum

- Lack of knowledge cannot be a result of a failure to expose students to the basic multiplication facts since they appear in more than one grade level in the authorized textbooks that are used. What may need to be examined is whether or not a program that is used devotes the necessary time to the pre-requisites that are required.

- Lack of knowledge cannot be a result of lack of practice, because the majority of the references that are used include many suggestions for practice. Perhaps the type of practice that is provided requires examination. Does the program provide rote or *appropriate practice* which contributes to the development of *numerical power*?

- In most cases lack of knowledge cannot be blamed on lack of long term memory since many students who are unable to recall basic multiplication facts know literally hundreds of songs from their favourite rappers, singers or groups of singers by heart. What might be done to reach these students?

- During diagnostic interviews some subjects are encountered who do not know the basic multiplication facts and they indicate that they did not want to know these facts. Since lack of knowledge of the basic multiplication fact means not knowing the basic division facts, these students face a handicap that they may not be able to overcome. Preparing an appropriate intervention **IEP** for these subjects can present a challenge.

Possible Specific Contributing Factors

- It is possible that a combination of factors come into play. During a visit to one school the request by a grade four teacher to teach the students strategies for the multiplication facts was unsuccessful because the students lacked certain aspects of *number sense*. The students were unable to *visualize* one-half and doubles of numbers that are important for the basic multiplication facts. *Number sense* is a pre-requisite. Without *number sense* the task of having students develop *mental mathematics strategies* for teaching themselves or learning the basic multiplication facts becomes impossible. Without *number sense* students are unable to make the *connections* and every fact has to be learned in isolation, a formidable task for some students.

- Displaying multiplication tables on students' desks and a focus on quick recall are detrimental to having students develop *mental mathematics strategies* for the basic multiplication facts. During one diagnostic session a student in grade five answered the question, *What do you do when you do not know or you forget the answer for a basic multiplication fact?* with, *I look it up on a chart.* When asked what he would do if a chart was not available, he stated, *I leave it blank.*

- Disagreement about definitions for *times tables* or *multiplication tables* can be confusing and lead to incorrect answers when mental mathematics strategies are employed.
  For example, which of the two, **A** or **B**, is the *Nine Times Table*?
  A) 1 x 9 = ☐ 2 x 9 = ☐ ... **B) 9 x 1 = ☐ 9 x 2 = ☐ ...**
  In any group of adults, and that includes teachers, some will select **a x 9** and others **9 x a,** and sadly a few will say that it does not make any difference since they both are the same.

  The inconsistent interpretations of **a x b** can lead to difficulties.(1) During diagnostic interviews, students have responded to requests like,
  *If **5 x 8 = 40**, how could you use that answer to find the answer for **6 x 8** = ☐?*
  with, *'Oh, that is easy, just add **6** to **40**.'* (2)

  During an interview a student from the fifth grade was asked what he would do if he did not know the answer for a multiplication fact. He responded with a comment that is music to any teacher's ears, *'I work on it.'*

  The student used the answers for five times a number (**5 x a**) and nine times a number (**9 x a**) as *benchmarks* and depending on the equation, skip counted up or down to arrive at the product. The subject referred to these *benchmarks* as the *'five times table'* and the*' nine times table'*. *This* is opposite of how these tables are defined in the curriculum. This confusion resulted in the addition of the incorrect factor when answers were calculated.
  For example, for **7 x 8 =** ☐ his explanation was,
  'I would go nine times seven and minus it by eight.'
  The response indicated that the subject is able to think mathematically, but a lack of *conceptual understanding* of multiplication leads to confusion and mistakes.

- The terminology that is used in some references and as a result in some classrooms does not foster *visualization* nor contribute to fostering *conceptual understanding* of multiplication. Examples of such terminology include: *times* and *multiplied by* for multiplication and *divided by*, *into* and *goes into* for division. Appropriate language enhances *visualization* and in turn can contribute to the learning of the basic multiplication facts.

Key Ideas from the Previous Grades

Fostering the development of *number sense* is an important component of the mathematics program in the primary grades. Specific learning outcomes from the primary grades dealt with the development of *conceptual understanding* of the four operations and the acquisition of *mental mathematics strategies* for the *basic addition facts* and *basic subtraction facts*. These key ideas from the previous grades can be used as part of pre-assessment, as part of diagnostic conversations or interviews and for content of introductory lessons.

Conceptual Understanding of the Four Operations

Important indicators of *conceptual understanding* include:

- The ability to listen to multiplicative action stories that are told or to problems that are read and translate these into equations. Appropriate equations identify the correct order and meaning of the numbers in the stories or problems.

- The ability to listen to multiplicative action stories or problems and draw appropriate sketches or diagrams that illustrate the order and the meaning of the numbers as well as the action that is described.

- The ability to translate sketches or diagrams that illustrate multiplicative actions into appropriate equations and word problems from their experiences that illustrate the meanings and the order of the numbers.

- The ability to match equations with appropriate problems from settings outside the classroom and to illustrate the numbers, the action and the correct order with action sketches.

  a + b = ☐   a – b = ☐   a x b = ☐   a ÷ b = ☐

  Addition: Does the problem make reference to the additive action? Is the order of the numbers correct? Is the action illustrated?

  Subtraction: Does the problem make reference to the subtractive action? Is the action correctly illustrated?

  Multiplication: Is the correct meaning for each numeral used in the problem and sketch, i.e., the first numeral tells how many groups there are and the second how many there are in each group?

  Division: Is there awareness that there are two different meanings that can be assigned to an equation; *equal grouping* or *taking away groups of* and *equal sharing*. For example, **6 ÷ 2 =** ☐ could be an equation for,

  How many teams of two players? or,

  How many players on each of two teams?

  Each of these problems has a unique solution procedure.

Knowledge of Basic Addition and Subtraction Facts

Indicators of knowledge of *basic addition* facts include:

- The ability to explain, in ways other than counting, that the answers for given equations are correct. Indicators of *number sense* will become apparent when students can describe more than one *mental mathematics strategy* to meet this request.

- The ability to explain strategies that could be used to teach someone who had forgotten some of the answers to the basic facts.

- The ability to make up and apply personal rules for recording the answers for equations that have zero or one as an addend.

Indicators of knowledge of basic subtraction facts include:

- The ability to explain that every subtraction equation can be solved by connecting it to a known answer for an addition equation.

- The ability to make up and apply personal rules for subtracting zero from a number; one from a number; and a number from itself.

Number Sense

The indicators of *number sense* that are essential for having students develop *mental mathematics strategies* for the *basic multiplication facts* include:

- The ability to *visualize* and state the least number of base ten blocks or the least number of students it would take to show the number for a two digit numeral with their fingers.

- The ability to *visualize* and name one-half of numerals that are presented.

- The ability to *visualize* and name the double for numerals that are presented.

Indicators of *number sense*, or lack thereof, can become evident when the request is made to show and to explain how fingers could be used to calculate the answers for given addition and subtraction equations. For example:

7 + 9 = □ **13 – 7 = □**

Counting up by ones from the first addend and counting down by ones from the minuend are indicative of the fact that students are unable to visualize and think of different names for numbers.

The counting strategy subjects use to explain how they know that a given equation is not true can yield information about *number sense*. For the example **7 + 9 = 15** some subjects who were interviewed began by drawing seven dots. After drawing nine more dots, all of the dots are counted to determine the total.

During interviews some subjects will begin to show how to use their fingers to calculate the answer for **13 – 7 = □** by making reference to counting imaginary fingers. Sometimes two numerals are named for each separate step of the count, i.e., *'Twelve – one, eleven – two, ten – three, ... , six – seven'* to arrive at the answer; a rote procedure that can be very taxing on short term memory. Any distraction or interruption and the whole cumbersome procedure has to be started over again.

Teaching about the Basic Multiplication and Division Facts

The phrase 'teaching about' is used in the title because it can be argued that it is impossible to teach the basic facts to students. The most important goal for teaching about the basic facts is related to providing opportunities that will enable students to develop *mental mathematics strategies*. These strategies make it possible to have students teach the basic facts to themselves. These strategies also make it possible for students to get themselves unstuck should a basic fact be forgotten. This goal of teaching about the basic facts implies that settings, activities and problems need to be presented for reflection and the development of strategies and it rules out tasks that focus on producing answers quickly.

The Basic Multiplication Facts

An introduction to the one hundred basic multiplication facts, or all possible single digit combinations, needs to tap students' *sense of number* and their *conceptual understanding* of multiplication. As part of this teaching, there needs to be an emphasis on *connecting* or transfer from what is known. The 'groups of' interpretation is well suited to make connections to previous learning. It is easy to visualize settings that lack these characteristics. An introduction that begins with an activity that has students draw groups of dots and then counting the dots to determine the products is an example of a setting that is rote procedural.

The file of diagnostic interviews includes the following conversation with a subject who had almost completed grade four.
His response to, *How would you calculate the answer for 3 x 9?* was,

I think of nine circles with three dots, keep in mind
what it adds up to and know how many to go.
Sometimes I'm too much or not enough.
I get it wrong because I am not a computer.
I always make the biggest number into circles.
His response to, *Why?*
I like the lowest number to be dots. I just do.

These responses give an indication why the subject was referred for an interview. It is also possible to make a guess about the setting that was used as an introduction to the basic multiplication facts.

Mental Mathematics Strategies for Lists of Multiples

The development of *mental mathematics strategies* is well suited for *through* or *via* problem solving settings. The description of examples that are described and the discussion will make reference to *lists of multiples* rather than *times tables* or *multiplication tables*.

Doubling and Halving – Generating Lists of Multiples

Two equations from a list of multiples are presented. For example, for a list of multiples for four (**a x 4**):

6 x 4 = 24   2 x 4 = 8

The request is made to try and think of ways of using the given equation to calculate all the missing multiples:

from **1 x a =** ☐   **2 x a =** ☐   **3 x a =** ☐ ... to **10 x a =** ☐

and to be ready to share the strategies that were used to record the missing products.

The intent is that halving and doubling are used to go from **4 x a =** ☐ to:

2 x a = ☐   **1 x a =** ☐   **8 x a =** ☐

The additive property is then used to record the products for:

3 x a = ☐   **5 x a =** ☐   **6 x a =** ☐   **7 x a =** ☐   **9 x a =** ☐   **10 x a =** ☐.

The order of recording the equations may differ, but the strategies that will be used, discussed and compared will be similar.

The activity illustrates the importance of key aspects of *number sense* as necessary pre-requisites: *flexible thinking* about numbers and being able to *visualize* the results of determining one-half of a number and doubling a number.

Types of Activities and Problems for Lists of Multiples

Follow-up activities for lists of multiples that have been generated can include:

- Arranging the list of multiples that has been generated from the least to the greatest product. Predicting the next three or four members of the list. Explaining, orally or in writing, the strategy used to make the predictions

- Examining the products of a list to see if there is something the same about the numerals in the products. Is it possible to identify a property that is the same for all the members of the list of multiples?

- Highlighting the products of a list of multiples on a **99-Chart**. Is there evidence of a pattern? If there is a pattern, how could it be described and how would it help in identifying the next four members of the list?

- Presenting lists of different sets of numerals, one list at a time, and presenting the challenge of attempting to identify the numerals thought to be part of a given list of multiples highlighted on the **99-Chart**. Explaining the strategy that was used.
  For example,

  Which numerals are multiples of four? How do you know?

  8 10 12 14 16 18   14 16 20 22 24 28 34

- The **x** sign is assigned a meaning of *jumps from* or *steps from* zero. For example, **3 x 4** is thought of as *three jumps of **4** from zero.* A highlighted numeral from the chart is announced, i.e., *thirty-two.* The challenge to make a guess and state how many jumps it would take to get from zero to thirty-two is presented. This type of request could be preceded with an invitation like, *Thumbs* up *if you think more than five jumps or thumbs down if you think fewer than five steps?*

  The request can be made to show with fingers the number of jumps it takes to get to a multiple that is announced. The guesses made during this early type of short term memory task should eventually result in the ability to *visualize* the number of jumps on the chart as the activity is repeated over time. The goal is to reach the stage of showing the correct number of fingers or jumps for any announced product from different lists of multiples.

- After an example of a riddle for a mystery product is examined, the request is made to write a riddle about a member of a list of multiples that is highlighted on a chart. The riddles are shared.

What is the mystery product?
To find the name of the number from the chart, start at **0**
and take jumps of four.
After each hint explain what you know about the mystery product.
It takes fewer than 8 steps of 4 to find it.
It takes more than 2 steps of 4 to find it.
The sum of the two digits of the multiple is greater than 5.
The difference of the two digits of the multiple is not equal to 2.

- The request is made to explain, in writing, how the answer for five jumps, i.e., **5 x 4 = 20**, could be used to record the answer for other questions about how many jumps it takes. The explanations are shared and compared.

- Other types of problems:

How would you find a numeral that is a multiple of **6** and **7**?
Explain your strategy.
How is it possible to find numerals that are multiples of **2**, **3**, **4**, **6**, and **8**?
Try to think of at least two different ways.
Explain your strategies?

Skip Counting From a Benchmark

Skip counting as a *mental mathematics strategy* has to begin at a *benchmark*. The two benchmarks are **10 x a =** □ and **5 x a =** □. After several equations of the type **10 x a =** □ are solved, the request is made to make up a rule for recording the products for ten times a single digit numeral.

- **Accommodating Responses**
  There will be students whose rule will make reference to *adding zero to the single digit numeral*. This response provides an opportunity to mention and illustrate with examples the importance of using correct language.

The request is made to explain how the products for **10 x a =** □ can be used to obtain the answers for **5 x a =** □. Then the requests are made to explain how the answers for:

4 x a = □ **3 x a =** □ **6 x a =** □ **7 x a =** □ **9 x a =** □ **8 x a =** □

can be calculated from knowing the answers for **5 x a =** □ and **10 x a =** □.

Simple Properties

The requests are made to make up rules for recording products when one of the factors is **0** or **1** and to test or apply the rules. Time needs to be set aside to reflect about these properties. Responses collected during diagnostic interviews from subjects who experience difficulties show that these properties are confused, especially for multiplication and division.

The following is an excerpt from a diagnostic interview.[2] After printing the answers **4**, **3**, and **8** for **4 x 0 =** □, **0 x 3 =** □ and **8 x 0 =** □ respectively, one girl created the rule,

Well, zero doesn't work.
You can't times a zero by four or else it would have been zero
so you put the four down.
So you look at the next one and you can't put the zero
because then it would be wrong,
so you put the three, that's how it works.

The delivery and intonation of the last part of the explanation was interpreted as a definite indicator of,

Do not ask me any more questions about this.

Two other things are noteworthy. The girl's answer is incorrect, but the delivery was used as an indicator of being willing to talk with confidence in one's own natural language. This is a favourable characteristic for an intervention setting. Many students who are interviewed will use times as a verb. Responses are made in the form, *I timesed it by*. The expressions, *I plused* and *I minused* also appear during conversations.

About 'Tricks' and Patterns

During diagnostic interviews that probe insight into knowledge of basic multiplication facts, responses from subjects are encountered that make reference to tricks they have learned. Some subjects actually use trick as part of their explanations.

If by chance students are shown something that is labelled a trick, they need to be challenged to try and find out how and why a so-called trick works. They need to learn that tricks do not exist in mathematics.

The most frequently encountered 'trick' subjects will share is related to the use of fingers to calculate the products for the list of multiples of nine. Transcripts from interviews show that not one subject was able to explain how and why this 'trick' works.

Subjects have also been interviewed who know a 'trick' of how to record the answers for equations with factors greater than five and less than eleven with the aid of their fingers without being able to explain the mathematics behind the 'trick.' Knowing the mathematics behind this procedure would enable students to use and extend this method to other pairs of factors, i.e., for factors **11** to **15**.

There are some who claim they can teach the basic multiplication facts by looking at patterns. Students with such a background may not find it easy to provide answers for facts that appear in isolation. A heavy emphasis or reliance on patterns can be detrimental to the development of *mental mathematics strategies*.

Basic Division Facts

The equal grouping interpretation of division makes it possible to connect knowledge of the basic multiplication facts to the basic division facts. Every basic division fact equation can be solved by thinking of a matching multiplication equation.
For example,

42 ÷ 6 = ☐ can be thought of as,
*How many groups of **6** are equal to **42**?*

Interpreting the division equation or reading it as,
*What is the total number of groups of **6** that can be taken from **42?***
contributes to fostering *visualization* of the procedure and enhances the connection to multiplication. The inclusion of the reminder *total* or *taking away groups until all are gone* as part of the interpretation can assist with arriving at a generalization for division by zero.

The familiar and commonly used expressions *'divided by'* or, *'goes into'*, which has been labelled mathematical slang by some authors[3], or simply, *'into'* do not contribute to any *visualization* and transfer. In fact, responses from subjects during interviews show the latter two can be and are very detrimental. These expressions along with the symbol $\overline{)\quad}$ lead subjects to conclude that both *'two into six'* and *'six into two'* are appropriate interpretations for **6 ÷ 2 = ☐**.

Initial activities with basic division fact equations allow for a guess and test approach. After an equation is briefly shown, an initial reaction can be requested in terms of the benchmark *five groups* or *five multiples* of a number.
For example, for **56 ÷ 8 = ☐** the requests could be,

Do you think there are more than or fewer than five groups of eight?
How did you decide?
How do you know the answer is not six groups of eight?
What is the answer? How do you know it is the correct answer?

Simple Properties

The requests are made to make up and test rules for calculating the answers when:

- Dividing a number by one; **a ÷ 1 = ☐.**
- Dividing a number by itself; **a ÷ a = ☐**.
- Dividing zero by a number; **0 ÷ a = ☐**.

Division by zero requires special attention. During visits in classrooms, there have been times when the response to the request to tell something about an equation of the type **a ÷ 0 = ☐** is met with comments that make reference to the fact that,

You cannot do it because you cannot take away groups of zero from a number.

It is simple to simulate the action that shows that it is possible to take away groups of zero from any number, but that something else is impossible.
For example,

A simulation of the divisive action for an equation like **4 ÷ 0 = ☐** can include recording checkmarks on the chalkboard every time the action for taking away a group of zero is simulated. This could be continued until the majority of the students use their own expressions to describe reactions to the task.
The following questions could be posed,

*What answers might people be tempted to record for **4 ÷ 0 = ☐**?*
How do you know these answers are incorrect?
What will you think or say every time you see an equation of this type?

The following is an excerpt from a diagnostic interview with a student from a middle school. The subject recorded the following answers:

a ÷ 1 = a **a ÷ a = 1** **0 ÷ a = 0** **a ÷ 0 = 0**

After he attributed knowing the answers to his parents, the conversation continued as follows:

Have you ever heard that from your teacher?
Um, we don't do zeroes. They're too easy.
Who says they're too easy?
Teacher says we should know them.
What about teachers from a few years ago?
Only in grade four. We did about two pages and the first had about three questions on it and it went to fours – from dividing by zero and skipping to dividing by fours – because the ones and twos were too easy.

What an amazing memory! Even if the memory is questioned, many students and many adults provide similar statements as part of their responses to questions about division by zero.

Order of Operations

The ideas that relate to order of operations are well suited for teaching *through* problem solving type settings while working with a partner.

Equations, requests and questions of the following type could be presented:

a) **2 + 4 x 3 = ☐** **7 – 3 – 2 = ☐**

Use each numeral and operation only once and try to come up with two different answers.
If you think one answer is correct, which one do you think it is?
Why is that the case?
If you think both answers are correct, what problem or problems could occur?
What do you think could be done so everyone in the classroom and in the school will record and agree on the same correct answer?

b) **16 ÷ 4 – 2 x 2 + 4 = ☐**

The request could be to try and come up with as many different answers as possible or to think of ways of grouping the numerals to come up with the answers:

4; **12**; **16** and **24**.

To make it possible to keep a record of the strategies and to show how numerals are grouped, the notion of using parentheses is introduced. Letters of the alphabet above each grouping can be used to indicate the order of the procedure that was employed to arrive at the different answers.

Before the rule for order of operations is introduced, that is:

multiplication and division are done first from left to right,
and then addition and subtraction;

problems of the following type could be solved:[4]

Use grouping to make the equation true: **11 – 5 x 2 + 3 = 30**

Use grouping to get the answers: **4 + 8 ÷ 4 – 2 = 6** **4 + 8 ÷ 4 – 2 = 1**

Determine the appropriate operations and use grouping to make the equation true:

5 ⌑ 4 ⌑ 2 = 2

Assessment Suggestions

The written or oral responses provided by students can yield indicators of:

- Presence of *mental mathematics strategies.*
- Key aspects of *number sense – visualization* and *flexible thinking* about numbers.

Some responses will provide information about students' ability to *generalize.* Indicators of *confidence* and ability to talk about what has been learned in one's own words will also become available.

Collecting Data about Knowledge of Basic Multiplication Facts

- Pretend you forgot the answers for **7 x 8 = ☐**.
  How could you calculate the answer or how could you teach someone how to calculate the answer? Write a sentence that explains your thinking.

- If you know that **5 x 7 = 35** shows the correct answer, how could you use that to calculate the answer for **6 x 7 = ☐**? Write a sentence that explains your thinking.

- Use the equation **2 x 6 = 12** to write as many equations for multiples of six as you can. Explain your thinking for each equation.

- Write a rule for calculating the answer when a number is multiplied by one.

- Write a rule for calculating the answer when a number is multiplied by zero.

- A sheet of **25** or more different equations representative of the basic multiplication facts is presented and the students are asked to underline the equations that they think they can record the answer for just by quickly looking at them. They are asked to be ready to explain the strategies they would use to record the answers for the remaining equations.

  A variation to the setting could have students look at a sheet of equations and use two different colours to identify equations the students think are very easy and those that they think are not so easy. A three way categorization could be explained: very easy; easy; and not so easy. The students are asked to explain how the decisions were made. The coloured responses can yield valuable information for a focus of future activities or for intervention tasks.

Collecting Data about Knowledge of Basic Division Facts

- Pretend the answer for **42 ÷ 6 = ☐** is forgotten. How would you calculate the answer or how would you teach someone to calculate the answer? Write a sentence that explains your thinking.

- How do you know the answer in the equation **72 ÷ 8 = 8** is wrong? Write a sentence.

- Why is calculating the answers for **9 ÷ 1 = ☐** and **13 ÷1 = ☐** easy? Write a sentence.

- Why is calculating the answers for **8 ÷ 8 = ☐** and **12 ÷ 12 = ☐** easy? Write a sentence.

- How do you know that the answer in the equation **0 ÷ 7 = 0** is correct? Write a sentence.

- How do you know that the answers in the following equations are incorrect?
  3 ÷ 0 = 0 **5 ÷ 0 = 5** **4 ÷ 0 = 1** Write a sentence.

- What would you say to someone who records a numeral as the answer for an equation like **9 ÷ 0 = ☐**? Write a sentence.

Reporting

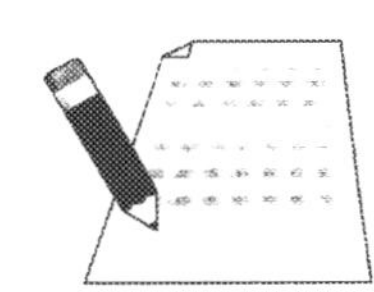

Many parents believe that the knowledge of the basic facts is one of the most important outcomes of mathematics learning. That is especially true for knowing the *basic multiplication facts*. Many parents believe students should be able to recite these facts quickly.

The assessment data that are collected can make it possible to share information about:

- Knowledge about what to do when an answer might be forgotten.
- Ability to explain how it is known that answers are correct.
- Ability to generalize about simple properties.
- Ability to identify answers that are incorrect.
- Indicators of number sense.
- Indicators of confidence and willingness to take risks.

Parents can be informed that as a result of the activities that were examined and the problems that were solved the answers to the basic facts are known, but more importantly, strategies have been learned that make it possible to reinvent the answers for a basic facts that may have be forgotten.

For Reflection

1) What are some possible disadvantages of having students look up the answers or check the answers for basic multiplication facts on tables that are taped to their desks?

2) What are some possible disadvantages to have students check the answers for basic multiplication facts on a calculator?

3) Why do you think many students and many adults think that the answers for **4 ÷ 0 =** ☐ and **0 ÷ 4 =** ☐ are the same?

Chapter 4 – Number: Computational Procedures (+, –, x , ÷) for Whole Numbers

Personal Strategies

Different determinants of the curriculum and the many aspects of mathematics learning make it impossible to describe specific characteristics of settings and teaching strategies that are true for every mathematics classroom, especially in North America. The references that are used by teachers may differ; teachers may have different backgrounds or they may have different beliefs about how mathematics is learned and should be taught. However, it is possible to describe specific characteristics of classroom settings and general strategies that are part of different methods of introducing students to computational procedures or algorithms.

The procedures that are used to record one's thinking as answers are calculated for items that involve numerals beyond the basic facts are labelled *algorithms*. The computational procedures adopted by a particular mathematics curriculum for the presentation in schools are sometimes referred to as the *standard algorithms*. Since different *algorithms* exist for each operation, the *standard algorithms* adopted by different countries as part of their mathematics curriculum are not always the same.

It is suggested in the curriculum that students be given the opportunity to develop *personal strategies* or *personal algorithms* for computational procedures. This method of learning differs from one where each step of a standard algorithm is demonstrated and then practiced. Since the development of personal strategies or the learning of computation procedures *through problem solving* differs from how most adults learned these procedures, it is deemed advantageous to make a comparison between the two instructional settings by illustrating the key characteristics of each one.

General Characteristics of Types of Classroom Settings

Teacher Directed Strategies – Standard Algorithms

The setting most adults are familiar with involves direct instruction to introduce computational procedures. The key parts of this type of classroom setting are demonstrations followed by repetitive practice. The sequence of presenting a computational procedure is based on a task analysis that determines the order of a hierarchy of skills. A specific example of such a hierarchy is included in the discussion about the teaching of the addition algorithm. For some computational procedures students repeat and memorize the exact phrases that were presented as part of the demonstrations. These types of settings can be described as being teacher-owned since a teacher takes the role as an inculcator and the student a rule taker.[1]

During diagnostic interviews subjects from direct instruction settings will recite the steps, or chants and dance steps as someone called them, that were memorized. These students are unable to state any reasons for the 'moves' they make with symbols on a piece of paper. Some will match a memorized chant with the wrong step, i.e., for 'invert and multiply' the wrong fraction is inverted.

Personal Strategies – Personal Algorithms

The major premise for settings that enables students to develop personal strategies for computation procedures is based on *sense making or on making mathematics meaningful.* Learning is *through problem solving* and a teacher takes the role of an educator.[1] There are those who react to this type of setting by sharing the observation that their students are not ready to 'invent mathematics'.[2] The invention of computational procedures is not a goal. The intent is to show students that everything they know and everything they have learned can be transferred to new procedures or skills.

There exist important pre- and co-requisites which enable students to develop *personal strategies* for computational procedures. These requisites include, first and foremost, *number sense* and *conceptual understanding* of the operations. Attempting to develop one's own computational strategies requires a degree of *confidence* and *willingness to take risks.*

Any group of students that lacks important requisites, especially those related to *number sense* and *conceptual understanding*, presents a formidable challenge for any teacher. A lack of these requisites will make it very tempting to move to a directed instructional setting with a focus on rote learning and repetition[2], much to the long term detriment of the students.

It is possible to obtain short term results on certain types of tests from settings that are based on direct instruction that focuses on: memorization of rules or steps; repetition; and rote practice. A good short term memory makes it possible to parrot what has been memorized in test settings that do not require any conceptual understanding, high order thinking or aspects of mathematical reasoning.

Frequently reports appear in newspapers about university students lacking basic understanding of mathematics or lacking survival skills. One university president made the statement that many of the mathematics majors know less mathematics than they think they do because they cannot talk about the mathematics they have learned in their own words. Over the years many students in education courses have made comments that agree with the writer of the letter to the editor who stated that, *his high marks on mathematics tests masked ignorance of math theory and understanding.*[3] Memorizing to pass a test is of no benefit. Being coached to do well on a test at best makes students test-wise but not *numerate.*

Enabling students to come up with *personal strategies* for computational procedures can have many positive outcomes. The experiences of such a setting help to make mathematics meaningful, facilitate future learning of mathematics and contribute to building *confidence* and *willingness to take risks* – essential components of the ability to solve problems.

Backgrounds of Students Arriving in the Intermediate Grades

It is possible that the students who arrive in the intermediate grades learned about the addition of numerals beyond the basic addition facts in a direct instruction setting.
A brief conversation, a part of a diagnostic interview or a few written responses to several key questions will provide the required information.

One type of diagnostic task consists of presenting an equation, i.e., **38 + 54 = □**, and requesting an explanation, orally or in writing, for how the answer is calculated.
Many of the responses to this request are predictable.
They start with,

First you write one number below the other.

Some students will state,

The biggest number goes on top.
Then you start with the ones.
You add the eight and the four that makes twelve, print the two and carry the one, etc.

The goals of a conversation or an interview are to probe understanding and thinking. Without knowing what students are thinking, it would not be possible to help them. As explanations are given more questions need to be asked.
For example,

Why do you write one number below the other?

Students who can give a meaningful reason for doing so are the exception. Anyone who has conducted interviews knows that at times responses can be quite amusing.
One subject responded to the last question with,

So I don't have to move my eyes so much.

Another student claimed that,

The teacher does it that way so she can get more questions on the blackboard.

Then there was the student who stated,

So I don't get confused.

When asked,

Confused with what?

the somewhat emphatic response was,

I don't know, but I don't get confused.

During conversations or interviews very few students make reference to the actual value of the digits they are manipulating or *'carrying.'* The majority of subjects use the term *carry*. Where did this term originate?

During conversations with students a lack of indicators of *number sense* can be noted. This lack is not surprising since it is likely due to the fact that *number sense* was not the focus of instruction for these students. Any student who during a conversation or an interview can provide reasons for the moves that are made with symbols is an exception.

The learning about the addition of numerals for students who arrive in the intermediate grades may have been based on a very detailed task analysis for the topic. Such an analysis identifies a hierarchy of skills. These skills are presented in order and students spend time practicing each new skill.

A task analysis for addition skills might look like this:

- A two-digit numeral plus a one-digit numeral with the sums in the ones place less than ten: **24 + 3 = □**
- A two-digit numeral plus a two-digit numeral with sums in each place value position less than ten: **24 + 32 = □**
- A two-digit numeral plus a one-digit numeral with a sum equal to ten: **24 + 6 = □**
- A two-digit numeral plus a one-digit numeral with a sum greater than ten: **24 + 8 = □**
- A two-digit numeral plus a two-digit numeral with the sum in the ones place greater than ten: **24 + 39 = □**
- A two-digit numeral plus a two-digit numeral with the sum in the tens place greater than ten: **24 + 93 = □**
- A two-digit numeral plus a two-digit numeral with the sum in the ones and the tens place greater than ten: **24 + 98 = □**

Such a hierarchy of skills makes it easy to visualize the sequence of demonstrations during directed teaching and the examples of repetitive practice. For the next unit of the topic or the next grade level, the hierarchy is repeated as new place value positions are added. These kinds of experiences do not lead to any generalizations nor do they contribute to the development of *number sense* or to any kind of *mathematical reasoning* or *willingness to take risks.* These results are illustrated repeatedly by the responses collected from students during diagnostic interviews.

Teaching about Addition in the Intermediate Grades

The discussion about the teaching about addition is guided by the following questions:

- What major conclusions about addition will students reach as a result of the development and employment of *personal strategies*?
- What skills and procedures should students learn about the addition of three- and four-digit numerals?
- What should students be able to conclude about adding numerals in columns?
- What are some possible appropriate practice activities and problems?

Key Ideas from the Previous Grades

Conceptual Understanding

Conceptual understanding of addition enables students to make up meaningful word problems from their experience for given equations, i.e., **9 + 8 = □**; **36 + 42 = □,** that demonstrate an understanding of the numbers behind the numerals, the order of the numbers and the additive action.

Basic Facts

Knowledge of the basic addition facts implies the ability to use at least two *mental mathematics strategies* to explain why answers are correct, i.e., **9 + 8 = 17** and to use *mental mathematics strategies* to re-invent an answer for any basic addition fact that may have been forgotten.

Number Sense

Indicators of *visualizing* numbers include the abilities to suggest:

- The least number of students it would take to show the numbers for any two-digit numeral with fingers.
- The least number of base ten blocks or bills of denominations **$1 000**, **$100**, **$10** and **$1** (Loonies) it would take to show the numbers for any three-digit numeral.

The ability to *think flexibly* about numbers implies knowing that numbers can be represented and named in different ways and knowing the results of trading one ten for ten ones; one hundred for ten tens; and one thousand for ten hundreds.

Personal Strategies

Students know and can explain in their own words at least one procedure for calculating the answers when two two-digit numerals are added; or a two-digit numeral is added to a three-digit numeral.

Addition with Three- and Four-Digit Numerals

The addition of two or more multi-digit numerals involves two main generalizations. Digits in the same place value position need to be kept track of when numerals are added and anytime the sum in a place value position is equal to ten a group of ten is added to the next highest place value position.

It is possible to use **examples** as well as **non-examples** as part of demonstrations that can lead students to draw conclusions and arrive at the desired generalizations. The same goal can be reached by posing an **open-ended question** or problem.

Keeping Track of Digits in the Same Place Value Positions

Examples:
The request is made to observe as the following calculations are performed on the chalkboard.

$345 + 621 = \square$

345	345	345	345
+ 621	+ 621	+ 621	+ 621
900	60	6	6
60	900	900	60
6	6	60	900
966	966	966	966

Responses to questions of the following type are invited,

How do the calculation procedures differ?
How are the calculation procedures the same?
Which of the procedures do you prefer and why?

Two examples with five and six digits are presented, i.e.,

64 371 + 23 518 = ☐ **328 541 + 666 458 = ☐**

The request is made to try each example in two different ways and to record an answer to the question,

If younger students needed to learn one important idea about adding two numerals with many digits, what would that be?
Why is the idea important?

Non-examples:
The students are asked to observe as the following calculations are performed on the chalkboard.

345 + 621 = ☐

$$\begin{array}{r} 231 \\ +\ 152 \\ \hline 2462 \end{array} \qquad \begin{array}{r} 231 \\ +\ 152 \\ \hline 23252 \end{array} \qquad \begin{array}{r} 231 \\ +\ 152 \\ \hline 1751 \end{array}$$

Questions of the following type are posed,

What do you think is done incorrectly? Why is it wrong?
If younger students needed to learn one important idea about adding two numerals with many digits, what would that be?
Why is the idea important?

Open-ended Question
Pairs of students are requested to try to think of at least two different ways of calculating the answer for **634 + 255 = ☐** and to be ready to share explanations for the questions,

How do you know that the answer is correct?
How would you teach a younger student to add these numerals?

The methods: **examples**, **non-examples** and **open-ended questions** lead students to conclude that digits in the same place value position have to be added and an easy way to guarantee that this happens is to record numerals below one another and align the place value positions.

Keeping Track: A Group and Groups of Ten

One or more of the three methods, **examples**, **non-examples** or **open-ended** question can be used to illustrate keeping track of a group of ten or groups of ten.

Examples:

$$\begin{array}{r} 1 \\ 365 \\ +\ 492 \\ \hline 857 \end{array} \qquad \begin{array}{r} 1 \\ 58 \\ +\ 29 \\ \hline 87 \end{array}$$

Where did the one on top of the two examples come from?
What does each of the two ones mean?

Non-examples:

```
  68       593
+ 39     + 241
 917      7134
```

Where do you think the digits in these answers come from?
What mistakes do you think were made?
Why do you think anyone would make such a mistake?

Open-ended Question
Pairs of students are requested to use base ten blocks and show how to calculate the answers for equations like: **47 + 168 =** ☐ **256 + 174 =** ☐ and to keep a record on paper how the answer was calculated. The recordings are compared,

How are the recordings different?
How are the recordings the same?

Keeping Track of Groups of Ten

One of the *mental mathematics strategies* for the basic addition facts involves *going first to ten*, i.e., for **7 + 5 =** ☐ students think of **5** as **3** and **2** and then completing the equation by thinking **7 + 3 = 10** and two more is equal to **12**.
This strategy can easily be transferred to the addition of several numerals.
For example, **27 + 36 + 19 =** ☐

```
   2
  27
  36*
+ 19*
   2
```

Every time a ten is reached a tally mark is placed beside the appropriate numeral, in this case the **6** and then again beside the **9**. After **2** is recorded in the ones column, the tally marks are added in the appropriate column, in this case the tens column.

When the addends have three or more digits, the tally or check marks can be placed above the appropriate place value column. Prior to the addition of the numerals in a column, the tallies are counted and the numeral is recorded.

Estimating Sums

At one time students learned a rote procedure for estimating answers for equations or to check the reasonableness of their answers for the four operations. This procedure involved the use of rules to change the numerals in equations to 'rounded' numerals and every student was expected to use the same procedure.

Since the development of *number sense* takes time and it is not the same for every student in a group, it is appropriate to let students choose the numerals that appeal to them as predictions about answers are made. The numerals they choose, or their 'nice' numerals as someone labelled them, can provide possible indicators of *number sense*. *Number sense* fosters the ability to estimate, and being *confident* and *willing to take risks* while *estimating* and discussing the results keeps contributing to the development of *number sense*.

Activities need to be designed to help students become aware of the fact that different numerals can be used to estimate answers. For example, for several equations that are presented the following requests are made,

What numerals would you use to estimate the answer for **642 + 275 = □**?
Do you think your estimate is greater than or less than the answer?
Explain your thinking.

As pairs of numerals are shared students will become aware of the fact that not everyone used the same numerals. Each *estimation* task that is attempted should be accompanied by the challenge to try and tell whether or not students think the actual answer is greater than or less than the *estimate* and to be ready to explain the thinking that was part of the decision. This valuable aspect of thinking is an important part of problem solving since many problems require *overestimation*.

Mental Mathematics

Estimating enables students to make predictions about answers or check the reasonableness of answers. *Mental mathematics* enables students to calculate answers without the use of pencil and paper.

When requests are made during interviews to describe how answers could be calculated without the use of pencil and paper, many subjects will describe the exact steps they would use for the standard algorithm on piece of paper. These responses do not provide any information about possible indicators of *flexible thinking* and *number sense*.

To foster *flexible thinking* students should be asked to try several different methods for calculating answers without pencil and paper.
For example, **2342 + 449 = □**
Try to describe to your partner how you would calculate the answer in each of the following ways:

- *Start with the hundreds; then think about the thousands; then think about the tens; and finally think about the ones.*
- *Start with the tens; then think about the ones; then think about the hundreds; and finally think about the thousands.*
- *Start with the tens; then think about the thousands; then think about the ones; and finally think about the hundreds.*

The main idea of the setting is to encourage the use of several different starting points for the same equations. During follow-up discussions it can be determined whether or not there exists a consensus about preferred starting points or preferred sequences.

Appropriate Practice

Appropriate practice settings contribute to the understanding of the skills, procedures and ideas that have been examined. Opportunities are provided to use aspects of *mathematical reasoning*. The activities and problems contribute to fostering the development of *number sense*.

Types of Activities and Problems

Missing Numerals
Opportunities are provided to discuss and compare the strategies used to determine and record the missing numerals for items of the following type,

$$\begin{array}{r} 631 \\ +\ \square\square\square \\ \hline 987 \end{array} \qquad \begin{array}{r} 8\square \\ +\ \square\square 2 \\ \hline 305 \end{array}$$

The activities can include items that have more than one possible answer.

What numerals do you think are possible?
Explain your thinking.

$$\begin{array}{r} 3\square \\ +\ 4\square \\ \hline 84 \end{array}$$

An item can be included that is not possible. The intent is not to trick the students, but to see who is able to notice that something is impossible.

$$\begin{array}{r} 56 \\ +\ 2\square \\ \hline 73 \end{array}$$

A request should be made to keep a record of the different numerals that were used for the different attempts to identify the missing numerals. This record can be referred to as strategies are explained. The record also provides an indication about how many different equations were tried as part of the attempt to determine a possible solution.

Arranging Digits According to Instructions
An arrangement of boxes is provided. The numerals **0** to **9** are placed into an envelope. Five numerals are drawn or students working in pairs could list five of their favourite digits. For example,

$$\begin{array}{r} \square\square\square \\ +\ \square\square \\ \hline \end{array}$$

The five numerals drawn are: **7**, **2**, **4**, **8**, **3**.

Instructions are provided.
For example,

Use each digit once.

- Enter the digits into the boxes to obtain the greatest possible sum.
- Enter the digits into the boxes to get the least possible sum.
- Enter the digits into the boxes to obtain a sum between the greatest and the least sum.
- What are the different sums that are possible? How do you know you have all of the possibilities? Compare your examples with someone.
- *If you are able to repeat one of the digits, what would be the greatest sum and the least sum?*

For variations, the number of boxes, the arrangement of the boxes or both the number and arrangement can be changed.

Changing a Sum

After a sum has been calculated for an equation like **2352 + 483 = □**, requests of the following type are made,

- Think of at least four different ways to increase the sum by one hundred. *Record your methods.*
- *Think of at least four different ways to increase the sum by one thousand. Record your methods.*

Using a Calculator

Different Ways: Several equations are presented. The request is made to use a calculator and to calculate the answers for each of the equations in at least three different ways and to keep a record of the order in which the numerals were entered. This record keeping is important for two main reasons. Students can look at what they have recorded as they explain the strategies that were used. Without a record it is impossible to pinpoint any sources of possible mistakes that may have been made. It has happened that students who came up with an incorrect answer shared comments like,

I did exactly what you told me to do.
I used the same numbers but got a different answer.

Without a record of some sort, it is impossible to be of assistance. That is the main reason the calculators that are used in an intervention setting or as part of learning assistance require print capabilities.

Guess the Addends: Several items are presented for inspection. For example,

a)	b)	c)	d)	e)	f)	g)	h)
986	555	468	805	444	421	734	456
+ 347	+ 350	+ 811	+ 284	+ 644	+ 888	+ 188	+ 452

Two addends are entered into a calculator. Hints about the sum shown on the calculator are provided, one at a time, to the students.
For example,

- *The sum is greater than one thousand. Which items can be crossed out? Explain your thinking.*
- *The sum is greater than one thousand three hundred. What can be crossed out? Explain your thinking.*
- *The sum is a name for an odd number. What can be crossed out? Explain your thinking.*
- *It is the greater of the remaining sums. What is the sum?*

After several examples of this type or with examples that showed three addends have been attempted, the students can work in pairs. Turns are taken. One gives hints about a sum that is shown on the calculator and the other tries to guess the addends that were entered.

Finding the Mistakes

Examples that contain different calculation and recording errors are presented. The request is made to try and identify what was done wrong in each calculation and to make a guess about why the error might have been made. The errors are corrected.

The request could be made to record a reaction to the questions,

Do you think you would ever make these types of mistakes?
Why or why not?

For example,

$\begin{array}{r} 63 \\ +\ 45 \\ \hline 18 \end{array}$	The student added the digits or used base ten blocks and then counted the blocks.
$\begin{array}{r} 458 \\ +\ 25 \\ \hline 908 \end{array}$	The student added 250 rather than 25
$\begin{array}{r} 2582 \\ +\ 391 \\ \hline 28173 \end{array}$	The student recorded the added hundreds in the thousands column.
$\begin{array}{r} 639 \\ +\ 295 \\ \hline 824 \end{array}$	The student did not keep track of the groups of ten in the tens and the hundreds column.

Subtraction with Three- and Four-Digit Numerals

During a session with teachers-to-be the point was made that for the sake of communication there are times when it is advantageous to know the mathematical label for each of the parts of an equation. After the example of addends and sum was mentioned, a subtraction equation was pointed to. The initial numeral was identified as the minuend, the second as the subtrahend and as the third part was pointed to one student very impulsively interrupted and announced, '*That must be the other end.*' After a good laugh, it was concluded that *other end* may in fact be more appropriate than difference, because if the question, *what's the difference?* is posed, or posed repeatedly, some students may get the impression that the person who is doing the asking does not care about the topic of subtraction.

Key Ideas from the Previous Grades

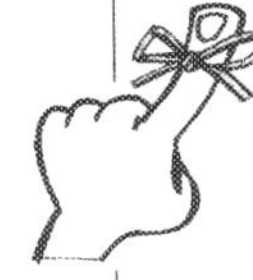

Conceptual Understanding

Conceptual understanding enables students to make up a meaningful word problem from their experience for given equations, i.e., **13 – 6 = ☐**; **52 – 27 = ☐**. The word problem demonstrates an understanding of the numbers behind the numerals, the order of the numbers and the subtractive action.

Basic Facts

Knowledge of the basic subtraction facts implies knowing the basic addition facts. Every basic subtraction fact can be derived from a known basic addition fact. This knowledge enables students to get unstuck should a basic subtraction fact be forgotten.

Number Sense

The key indicators of number sense are ability to *visualize* numbers and being able to *think flexibly* about them. The *flexible thinking* about numbers enables students to represent numbers in different ways and assign different names to numbers.

Personal Strategies

Students are able to demonstrate and explain at least one strategy for calculating the answer when a two-digit numeral is subtracted from a two-digit numeral.

Different Algorithmic Procedures for Subtraction

During the discussion about division it was pointed out that some authors label the expression *goes into* as mathematical slang. Perhaps the expression *borrow* should be classified in the same way. It is not a mathematical term and it does not make sense. It is fascinating to speculate how and why *borrow* along with *carry* first began to appear as part of mathematics teaching and learning.

The importance of keeping track of digits from the same place value positions is not something that needs to be part of teaching about a computational procedure for subtraction since it was part of the teaching about a computational procedure for addition. The most important idea that has to be part of activities and problems is enabling students to develop a strategy for dealing with situations when the digits in the subtrahend are greater than the digits in the minuend.

Different Computation Procedures

Different countries have adopted different *standard algorithms* for subtraction.

- Responses the subjects provide during diagnostic interviews will be indicative of the *standard algorithm* they were introduced to while learning a computational procedure for subtraction. For the majority of subjects an explanation for how the answer for the example could be calculated begins with a predictable recital.

$$\begin{array}{r} 52 \\ -27 \\ \hline \end{array}$$

  You can't take seven away from two, so you go to the five in the tens place. You make that four tens. Take the ten and add it to the two, that gives twelve, ... Some time ago, the mathematician turned entertainer, Tom Lehrer, actually made this type of chant a part of a popular song about the so called New Math which included the punch line, *the idea is to understand what you are doing rather than to get the right answer*. This is suggestive of the fact that at one time the ability to repeat a memorized chant was used as an indicator of understanding.

- The equal addition principle for subtraction can be used to avoid assigning another name to the minuend. The knowledge that adding the same number to the minuend and the subtrahend does not change the answer can, for example, result in changing an equation:

  from **52 – 27 = □** to **55 – 30 = □.**

  The answer can now easily be recorded.

Rather than adding a single digit numeral, the standard algorithm taught in some countries adds a group of ten to both the minuend and subtrahend where it is required.
For example,

$$\begin{array}{r} \\ 52 \\ \underline{-\,27} \end{array} \text{ is changed to } \begin{array}{r} 12 \\ 5\square \\ \underline{-\,37} \end{array} \qquad \begin{array}{r} \\ 341 \\ \underline{-\,89} \end{array} \text{ is changed to } \begin{array}{r} 11 \\ 34\square \\ \underline{-\,99} \end{array} \text{ and } \begin{array}{r} 14\;11 \\ 3\square\square \\ \underline{-\,199} \end{array}$$

- When subjects are asked during interviews to show how they would illustrate the calculation of an answer for a subtraction equation with base ten blocks, the majority of them will begin with the blocks that have greatest place value.
  One girl from the third grade used the following personal strategy to find the answer for **52 – 27 = □**.
  She took one ten from the five tens, returned three ones, added the ones and then subtracted the tens:
  $$\begin{array}{r} 52 \\ \underline{-\,27} \end{array} \text{ is changed to } \begin{array}{r} 42 \\ \underline{-\,2\square + 3} \\ 25 \end{array}$$

- Transcripts from interviews indicate that there are subjects who confirm that they have different ways of calculating answers. The subjects will make a comment like,
  With blocks I usually start with the big blocks, but on paper I do it differently.

- During a conference presentation one teacher shared a personal strategy that one of her students in grade five came up with. The student used the number line. For an example like **52 – 27 = □**, the student started at positive two, counted seven spaces to the left and ended up on negative five. Then he subtracted twenty from fifty and arrived at the answer by considering the net worth between thirty and negative five. This student does not require the chant that many subjects recite for this task during interviews,
  'You can't take two from seven so you go to the five in the tens place ...'

These examples illustrate that there are different algorithmic procedures that can be used to solve subtraction equations. When this is pointed out to teachers-to-be, many wonder out loud whether or not these strategies could be confusing to students. Without *number sense* that could indeed be the case. It is *number sense* that would enable students in the intermediate grades to understand each of these computational procedures. *Number sense* is the key for students being able to *transfer* what they know and *apply* it to equations that they have not encountered.

Open-ended Questions

Several equations are presented along with the challenge to try and think of different ways to calculate the answers.
For example,

342 – 71 = □   **621 – 88 = □**   **4444 – 555 =□**   **2000 – 199 = □**

The following questions could be part of this challenge:

- *What did you and your partner do to make it possible to calculate the answers?*
- *How could you use base ten blocks or money to explain how to calculate the answers?*
- *What would you teach younger students about calculating the answers for equations like this?*
- *What do the younger students need to know and do?*

The key generalization that is part of solving equations of this type is based on *number sense* or the ability to *think flexibly* about numbers. This generalization is the answer to the question, What numbers and number names make it easy to calculate the answer?

Estimating Differences

After an *estimate* is reported, the request is made to tell which numerals were used to arrive at the *estimate* and why these numerals were used. An explanation is requested for knowing whether or not an *estimate* is greater than or less than the actual answer.

3 986 – 294 = □ 317 – 58 = □ 6 243 – 169 = □ 936 – 297 = □

Mental Mathematics

As requests are made for calculating answers without the use of pencil and paper, the challenge of trying two different methods is presented. As methods of calculations are reported comparisons are made.

609 – 97 = □ 148 – 99 = □ 936 – 797 = □ 300 – 175 = □ 749 – 160 = □

Appropriate Practice

Practice settings should include opportunities to foster the development of *number sense* and *flexible thinking*. The sharing of strategies and solution procedures will contribute to the ability to communicate mathematically. Listening to others share their strategies can result in the discovery and possible adoption of new strategies.

Types of Activities and Problems

- **Missing Numerals**
  The tasks that are prepared should include items that have more than one possible answer. Once students discover that this is the case, they are asked to try and think of all of the possible numerals that can result in equations that are true.

  39□ – □7□ = 124 5□2 – □6□ = □23 □ □□□ – □22 = 4 222

  The tasks could also include examples that are not possible. For example,

  482 – 6□ = 423.

  Explain your thinking or the reason behind the decision.

- **Missing Numerals and Missing Signs**
  The initial decision involves choosing between addition and subtraction. Numerals are then chosen to write true equations and the selection strategies are shared.
  The following challenge is presented,

  Do you think it is possible to write true equations for the operation that was not chosen?
  If it is possible, what would the missing numerals be?
  Explain your thinking.

  64□ ⌑ □ □ □ = 321   **□□1□ ⌑ 1□4 = 841**

  The students are asked to keep a record of the numerals they tried as part of their attempts. These records can be used to compare strategies during a follow-up discussion.

- **Arranging Digits According to Instructions**
  An arrangement of boxes is provided. The numerals **1** to **9** are placed into an envelope. Five numerals are drawn or students working in pairs could list five of their favourite digits between **0** and **10**.
  For example,

  □□□   The five numerals drawn are: **9**, **4**, **7**, **3**, **1.**
  – □□

  Instructions are provided.
  For example,
  Use each digit once.
  - *Enter the digits into the boxes to obtain the greatest possible difference.*
  - *Enter the digits into the boxes to get the least possible difference.*
  - *Enter the digits into the boxes to obtain a difference between the greatest and the least difference.*
  - *What are the different answers that are possible? How do you know you have all of the possibilities? Compare your examples with someone.*
  - *If you are able to repeat one of the digits, what would the greatest and the least differences be?*

  After the completion of an example, students are invited to respond to,

  Explain your thinking about the two numerals when attempting to get the greatest (least) possible difference.

  The students can be invited to discuss the following scenario,

  If it is possible for one of the digits to be a zero where would you place it to get the greatest possible difference?
  Where would you place the zero to get the least possible difference?
  Explain your thinking.

  For variations to the setting, the number of boxes, or the arrangement of the boxes or both the number and the arrangement can be changed.

- **Changing a Difference**
  After the difference for an equation has been calculated, i.e., **3652 – 581 = □,** requests of the following type are made,
  - *Think of at least four different ways to increase the answer by one hundred. Record your methods.*
  - *Think of at least four different ways to increase the answer by one thousand. Record your methods.*

- **Using a Calculator**

Different Ways: Several equations are presented. The request is made to use a calculator and to calculate the answers for each of the equations in at least two different ways and to keep a record of the numerals that were entered into the calculator and the order in which it was done. Strategies are compared.

Guess the Minuend and Subtrahend: The setting that is described for having students **Guess the Addends** (p.91) for a sum that is shown on a calculator can be created for a list of subtraction equations.

- **Finding the Mistakes:**

What is wrong and why do you think the mistakes were made?
Do you think you would you ever make these mistakes?
Explain your answer.

359 – 167 = 192 **734 – 112 = 622** **289 – 198 = 91** **654 – 13 = 641**

Assessment Suggestions

The main purpose of the assessment is to gain insight into students' understanding of the computational procedures for addition and subtraction. The assessment strategies can be the same for both operations.

The types of assessment items that are presented consist of three-, four-, or a combination of three- and four-digit numerals.
Some of the assessment items that are presented could have more than one correct answer.

- Equations are presented, one at a time, or as separate items on a test.
  342 + 169 = □   651 – 275 = □   2000 – 359 = □
  Explain orally or in writing what you would do to calculate the answer.
  If you know of another way to calculate the answer, explain what it is.

- What do you think was done incorrectly? Explain or write a sentence.

$$\begin{array}{r} 2651 \\ +\ 5632 \\ \hline 7283 \end{array} \qquad \begin{array}{r} 367 \\ +\ 412 \\ \hline 4082 \end{array} \qquad \begin{array}{r} 4186 \\ -\ 269 \\ \hline 4123 \end{array} \qquad \begin{array}{r} 642 \\ -\ 199 \\ \hline 553 \end{array}$$

- Explain orally or in writing which numerals you would use to estimate the answers.
  Why would you use these numerals?
  Do you think your estimate will be greater than or less than the answer?
  How do you know?
  485 + 221 + 839 = □   6277 + 4885 = □

- What are the missing signs and missing numerals? How do you know?
  Explain the strategies that were used to calculate the missing information.

$$\begin{array}{r} 4\square9 \\ \square\ \ \square8\square \\ \hline 1\square8 \end{array} \qquad \begin{array}{r} 924 \\ \square\ \ \square5 \\ \hline 99 \end{array} \qquad \begin{array}{r} 753 \\ \square\ \ 4\square2 \\ \hline \square9\square \end{array}$$

Reporting

The responses on assessment items and the observations made during activities make it possible to generate various statements related to the understanding of the computational procedures. These statements make it possible to generate reports that include information about such ideas and skills as:

- Knowledge of more than one strategy to calculate answers.
- Ability to estimate to make a prediction about answers.
- Ability to explain whether or not estimates are greater than or less than the answers.
- Ability to use at least two mental mathematics strategies.
- Ability to identify errors in thinking, answers and/or use of materials.

The statements on reports can make reference to indicators of the presence of *number sense*, ability to *communicate mathematically*, *confidence* and *willingness to take risks*.
For example,

- Indicators about ability to *visualize* numbers noted as strategies were explained and illustrated with base ten blocks or with sketches.
- Indicators of ability to *think flexibly* about numbers and numerals observed when for a given equation students explained or illustrated:
  - two or more calculation procedures.
  - two or more mental mathematics strategies.
  - two estimation strategies.
  - two possible answers for missing numerals in an the equation.
  - two or more ways of using a calculator to solve the same equation.
- Indicators of ability to *communicate mathematically* and *confidence* noted as strategies were explained and shared.
- Indicators of *confidence*, *risk taking* and *perseverance* noted as attempts were made to find more than one possible solution for tasks.

For Reflection

1) What is a possible disadvantage of asking students to solve ten multi-digit addition equations and ten multi-digit subtraction equations and then request that a calculator is used to check the answers?

2) During a radio phone-in program a grade two teacher made the statement that at that time he was teaching his students *how to add three digit numbers.* What questions would you ask the teacher? Why would you ask these questions?

3) What would you say to a person who gives the following reason for asking a student to solve ten or more multi-digit addition items that are in a vertical alignment, *I want the student to practice the basic addition facts.*

Multiplication Beyond the Basic Multiplication Facts

At one time the task analysis used in the schools for teaching computation procedures for multiplication increased the digits in the multiplicand and the multiplier until practice items included two four-digit factors. The specific learning outcomes in the mathematics curriculum prescribe calculating the products for a three-digit multiplicand and a single digit multiplier and for two two-digit factors. It is safe to assume that a focus on *conceptual understanding*, *number sense* and appropriate practice will enable students to generalize and transfer the ideas, procedures and skills beyond these specific learning outcomes to multipliers with more digits.

Pre-requisites for Personal Computation Strategies

The ability to develop *personal strategies* for the multiplication algorithm is dependent upon several important pre-requisites. These include:

- The *groups of* interpretation for multiplication and knowing that in an equation the first numeral identifies the number of groups and the second numeral the number of members in each group. This interpretation makes it possible to visualize what is required to solve a given equation without ever having seen or attempted it before. For example, **12 x 24 = □** is thought of as finding the product for *twelve groups of twenty four*.

  Some students are capable of mathematical thinking but the terminology they are familiar with or use for the multiplication symbol results in confusion because this terminology does not foster *visualization*. This scenario is illustrated in Chapter 3 by the responses from a student in grade five who uses **5 x a = □** or **9 x a = □** as benchmarks and skip counts up or down to get himself unstuck, but then adds the wrong factors.

- Knowledge of the basic multiplication facts implies the ability to use *mental mathematics strategies* which include *halving* or *doubling a known fact* and then *skip counting up* or *down from the known fact* to calculate a basic fact or to reinvent a fact that may have been forgotten.

- *Generalizations* and rules about calculating products when numerals are multiplied by ten, by multiples of ten, by one hundred and by multiples of one hundred.

- Aspects of *number sense* that include the ability to *visualize* the numbers for two- and three-digit numerals and *think flexibly* about them.

- Knowledge of calculating the area for rectangular regions. This knowledge contributes to fostering the *visualization* of the partial products that are part of the computational procedure for multiplication and it accommodates making *connections*.

Results of Teacher Directed Settings – Indicators of Rote Procedural Learning

During diagnostic interviews indicators of having learned the computational procedures by rote become evident.[4] The computational procedure for calculating the product of two two-digit factors requires that the recording of the second partial product begins in the tens place value position. Many teachers request that for the second partial product a zero is recorded in the ones place. During interviews, the questions that probe possible reasons for placing a zero in the ones place has been answered with,

'I don't know, but it comes straight from my teacher's mouth.'
'I don't know, but my teacher tells me that I won't get mixed up.'

There are subjects who will come up with face saving comments during interviews. When requested to explain how to find the product for two two-digit factors, one subject declared,

'My teacher did this on the overhead. I didn't get it.
I sure hope we do it in grade five again.'

Then there was the subject who responded to the problem,

Five loggers. Each logger cut down twelve trees.
How many trees did the loggers cut down? with,
'Can you give me a minute?'

He drew sticks to represent the trees. After he had drawn the second group of twelve he uttered,

'I hope I don't have to do any more loggers.

When he was asked whether or not he knew of a faster way of sketching what was happening in the problem, he responded with, *'Sure'* and continued with the same routine, only at a much more rapid pace. He met the condition of the request, even if it was different from what was expected.

Other indicators of rote procedural learning that surface during diagnostic interviews include the referral of any pair of digits during the multiplication as ones and the product as ones and tens. It is fascinating to see how many students, and this is true for adults as well, will show a row of zeros in the second partial product when the multiplier has a zero in the tens place, i.e.,

142 x 103 = □.

Common reasons for the use of a row of zeros include phrases like:

'The zeros are place holders' and, *'The zeros help avoid confusion.'*

One student declared, *'Zeroes keep me in the rhythm.'* This response could be an indicator of steps or chants that were memorized as part of the calculation procedure.

Personal Computation Strategies

Open-ended Requests

The majority of pairs of students that have been challenged to try and think of at least two different ways to calculate the answer for an equation like **12 x 24 =** □ successfully met the request. This request to calculate the product was made prior to the introduction of any algorithmic procedure to the students and started with the presentation of a meaningful problem.
For example,

There are twelve cartons. Each carton contains twenty-four cans.
How many cans are there in the cartons?

The computational strategies the students came up with included:

- Calculating ten groups of twenty-four and then adding two more groups of twenty-four.
- Adding twenty-four twelve times.
- Calculating six groups of twenty-four and then doubling that product.
- Calculating two groups of twenty-four and then adding that product six times.

Generalizations

What are the main generalizations for calculating the products for two- and three-digit multiplicands and one-digit multipliers?

43 x 6 = □ **257 x 8 = □**?

Connecting equations of this type to problems of calculating the areas of rectangular regions enhances *visualization*. Sketches can be used to show the partial products that have to be accommodated as part of the calculation procedure.

For **43 x 6 = □** **6 x 40** and **6 x 3**

For **257 x 8 = □** **8 x 200** and **8 x 50** and **8 x 7**

Does it make a difference which partial product is first calculated?
Why or why not?

For **43 x 6 = □** **6 x 40 = □** or **6 x 3 = □**
For **257 x 8 = □** **8 x 200 = □** or **8 x 7 = □**

The possible recording procedures that indicate the order of thinking for the calculations can include:

43 x 6	43 x 6	257 x 8	257 x 8
18	240	56	1 600
240	18	400	400
		1 600	56

What other orders of calculations are possible for the three digit factor?
Which order do you prefer? Why is that the case?
Why is it advantageous to align the partial products as they are recorded?

To continue the fostering of *visualization* of numbers the request should be made to refer to the place value of the digits as the partial products are calculated.
The result of meeting this request for the examples above would be:

- *'Six times four tens is equal to twenty-four tens or two hundred forty.'*
- *'Eight times five tens is equal to forty tens or four hundred.'*
- *'Eight times two hundreds is equal to sixteen hundreds or one thousand six hundred.'*

If the products for one digit multipliers are recorded in one line rather than showing each partial product, a discussion is required about where to keep track of the digit that will be added to the following partial product. During the discussion possible advantages and disadvantages for each suggestion should be pointed out.

347
x 8
1 388

What are possible advantages and disadvantages of recording the **5** *tens and the* **3** *hundreds:*
- *beside the item?*
- *above the digit with the same place value?*
- *below the item in the appropriate place value position?*

What do you think is the best place for you? Why?

If guidance is required for the discussion, the following questions could be posed,

For which of these recording places might it be easy to forget the recorded digit?
For which of these recording places might someone be tempted to add the digit before the next multiplication is carried out?
What could be done to avoid using the same digit more than once?

- **Accommodating Responses:** The necessity of a brief follow up interview when something is done incorrectly[5] is illustrated by the following responses recorded by a student in grade five.

$$\begin{array}{r} 24 \\ \times\ 6 \\ \hline 1\,224 \end{array} \qquad \begin{array}{r} 33 \\ \times\ 4 \\ \hline 1\,212 \end{array} \qquad \begin{array}{r} 62 \\ \times\ 8 \\ \hline 1\,416 \end{array}$$

It may be tempting to draw a conclusion about what appears to be an obvious error pattern and to initiate intervention. However, when the student was asked to explain what he was thinking as he calculated the answers, he stated,

'I multiply six times four and that is equal to twenty-four.
I write down the four and carry the two tens.
Then I multiply six times two tens and that is equal to one hundred twenty tens.
Then I add the two tens and that gives me one hundred twenty two tens.'

The intervention for this response is very different than the one that is required for someone who records the partial products from right to left without doing the necessary renaming. This student's response was as surprising as the one from the grade three student who recorded the answer **25** for **32 – 17 = ☐**. The error pattern seemed obvious, but the student explained his thinking as,

'I can't take seven away from two, so I take ten from the thirty and add it to the two.
Now twelve take away seven is equal to five and thirty take away ten is equal to twenty.'

These examples provide more data in support of the statement by the authors[5] that more than fifty percent of the time our guesses about the possible reasons for the mistakes made by students may be wrong. Before students can be helped, it has to be determined how they think. In order to find out how students think, their explanations need to be listened to.

Visualizing the Partial Products

After a problem is presented and the matching equation has been recorded, the request is made to identify and name all of the partial products that have to be calculated. The partial products are listed.

There are fifteen boxes of apples. Each box contains thirty-six apples.
How many apples are there? **15 x 36 = ☐**

To solve the equation, the following partial products need to be calculated:

10 groups of **36** (**10 x 36**) or **10** groups of **30** (**10 x 30**) plus
10 groups of **6** (**10 x 6**) and,
5 groups of **36** (**5 x 36**) or **5** groups of **30** (**5 x 30**) plus
5 groups of **6** (**5 x 6**).

A rectangular region with dimensions **15** by **36** clearly shows all of the partial products that need to be calculated:

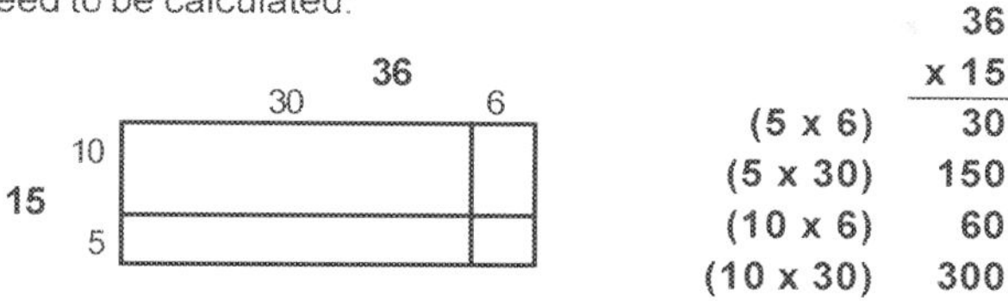

	36
	x 15
(5 x 6)	**30**
(5 x 30)	**150**
(10 x 6)	**60**
(10 x 30)	**300**

Is it necessary to reduce the number of partial products to two for calculating the product for two two-digit factors? A decision about an answer to the question has to give consideration to development of *number sense* and transfer. What are possible advantages of reducing the number of partial products that are recorded? Are there any disadvantages?

If it is deemed important and/or necessary to reduce the number of partial products, students should be requested to refer to the actual place value of the digits they are multiplying.

No matter what the algorithmic procedure, the goal for students should be to know why they make the moves they make with the symbols that are used and recorded. The actual meaning or value of each symbol that is manipulated should be stated. There should never be a reliance on so called short-cuts because these are labelled as being 'faster' by someone with many years of experience with these types of calculations. Short-cuts do not contribute to sense making or making mathematics meaningful.

The ability to *connect* equations to meaningful problems, knowing the meanings of the partial products of a calculation procedure, along with *confidence* and *willingness to take risks* will enable students to transfer the knowledge about multiplying two factors with two digits to personal calculation strategies for factors with more than two digits.

Appropriate Practice

The main goal of the practice items is to provide opportunities to *reason mathematically* as the computational procedures that have been introduced are examined.

Types of Activities and Problems

Another Algorithmic Procedure – Doubling
Number sense and *flexible thinking* about numbers will allow student to calculate products by using the doubling procedure or doubling algorithm used by Egyptians.[6] For this procedure one of the two factors is doubled until a combination of the other factor makes it possible to calculate the product. This procedure is related to the *mental mathematics strategies* students have developed for the basic multiplication facts.

For example, to calculate the product for **13 x 47 =** □ students can think and record,

1 x 47 = 47
2 x 47 = 94
4 x 47 = 188
8 x 47 = 376

Since **8 + 4 + 1 = 13**, the products for the factors **8**, **4** and **1** are added to result in the product for the equation,
376 + 188 + 47 = 611 or **13 x 47 = 611**.

The request can be made to calculate the products for several equations in two different ways.

How are the strategies the same and how are they different?
Which strategy do you prefer and why?

Missing Numerals

What strategies are used to identify the missing numerals? The request is made to keep a record of the attempts. The strategies are compared during a follow-up discussion.

The types of items that are presented can vary in levels of difficulty. Examples can be included that have more than one possible solution. An item or two could challenge students to discover that it is not possible to identify numerals that would result in an equation that is true.

□□□	□9	□4□	2□
x 9	x 7	x □	x □5
2007	35□	884	20
			100
			200
			□00

Arranging Digits According to Instructions

Four digits are listed, or students record their own favourite four digits, or they draw four digits from an envelope that contains the numerals **1** to **9**.

The request is made to draw an arrangement of boxes and enter the numerals into the boxes by following the instructions that are given.

For example,

a) □□□
x □

b) □□
x □□

*Use each of **2**, **3**, **4**, **5** once.*

Enter the digits into the arrangements of boxes to get:

- *The greatest possible product.*
- *The least possible product.*
- *A product between the greatest and least possible product.*
- *The greatest possible product that is a name for an even number.*
- *The least possible product that is a name for an odd number.*

The request is made to keep a record of the different entries that are tried for each request. This record is used as a reference as the strategies that were used are explained and compared.

Using a Calculator

Different Ways: The challenge is presented to try and calculate the product for an equation like **43 x 29 =** □ by entering numerals into a calculator in three different ways. A written record is to be kept of the numerals that were entered and the order they were entered. The methods that were used are compared during a follow-up discussion.

Guess the Factors: Several equations are presented for inspection.

For example,

a) 16 x 88 = □ **b) 20 x 30 =** □ **c) 13 x 13 =** □ **d) 16 x 47 =** □

One equation is entered into a calculator and as hints about the product are announced questions are posed.

For example,

The product is less than one thousand. Which factors could be crossed out? Why?

The product is a name for an even number. What could be crossed out? Why?

The sum of the digits in the product is less than ten? Which product is it? Why?

A list of equations is given to pairs of students. One student uses the calculator to obtain an answer for one of the equations. Hints about the product are presented to a partner who tries to identify the factors that were entered into the calculator.

Finding the Mistakes

What do you think is wrong? Why do you think the mistakes were made?

405	316	360	789
x 8	x 4	x 5	x 6
3 200	1 264	180	4 724

During the follow-up discussion the students compare the possible reasons they think might be responsible for the mistakes that were identified.

Estimating

How would you estimate the product for **43 x 27 = □**?

The focus of discussions about estimation strategies should be on two main ideas:

What numerals were used as part of the estimation strategy and why were these numerals used?

Do you think your estimate is greater than or less than the answer?

How was this decision made? Explain your thinking.

For one type of task, choices are presented and the request is made to identify the choice that would be selected to estimate the product for the equation and to explain the reason for the selection.

For example, for **88 x 13 = □** the possible choices could be:

a) **8** tens **x 1** ten or **80 x 10**
b) **9** tens **x 1** ten or **90 x 10**
c) **9** tens **x 2** tens or **90 x 20**
d) **8** tens **x 2** tens or **80 x 20**

Which of the choices for estimates **a)** *to* **d)** *do you think is greater than the product and which is less than the product?*

How did you make your decision?

For another type of task, equations and choices of estimates are presented.

For example,

42 x 36 = □ **160** **1 600** **16 000**.

The request is made to select the choice thought to be the best estimate and why that is the case.

As *number sense* develops, the range of choices for these type of items can be reduced.

For example,

from **1 000** **1 500** **1 800** to **1 200** **1 500** **1 600**.

The options of choices can be reduced to two choices.

For example, for **13 x 13 = □**

from **150** **200** **250** to **160** **210**.

During discussions about estimation, the students could work with a partner and try to calculate answers to questions of the following type:

Who do you think uses estimation strategies?

Try to think of some settings when estimation is used?

Why do you think that most of the time it is better to overestimate or to use an estimate greater than the actual answer?

Finding the Missing Product

A list of numerals is displayed.

752 442 552 222 453

It is announced that this list includes the product for an equation that is shown.

For example, **26 x 17 = □**

The students are challenged to identify the product without the use of pencil and paper.

Which choice or choices would you exclude? Why is that the case?

Which choice is selected? Why is that the case?

As the students examine examples of this type, the discussion could begin with,

Which of the choices listed do you think is obviously not the answer for the equation? Explain what makes it so obvious to you.

Several equations can be presented along with a list of products that presents more choices than there are equations.

Try to identify the products without any pencil and paper calculations.

Be ready to explain the strategies that you used.

28 x 16 = □			
42 x 39 = □	**2 898**	**5 070**	**448**
78 x 65 = □	**2 632**	**5 092**	**3 638**
94 x 28 = □	**1 638**	**1 522**	**2 048**

Finding the Missing Factors

A product and several equations are presented. The task consists of identifying the factors that will result in the product that is shown.

The product is **1 408,** *what are the factors?*

40 x 36 = □ **36 x 38 = □** **88 x 16 = □**

Explain your thinking.

Initially two of the choices given could make it easy for students to identify them as being inappropriate. This can then be reduced to one of the choices. As the ability to estimate develops, the range of the differences for the factors in the choices can be reduced. More than one pair of matching factors can be included in the choices.

Finding the Missing Factor

A multiplication equation with a missing factor is presented. The task consists of selecting the missing factor from several choices.

For example,

29 x □ = 1 218 **32 34 41 42 51 52**

A discussion could begin with,

Which of the choices listed do you think are obviously not the missing factors?

Explain what makes it obvious to you.

Which choice do you think is the missing factor?

Explain why you think that is the case.

Assessment Suggestions

The goal of the assessment is to gain insight into students' understanding of a computational procedure. A timed test setting that requires the calculation of products as quickly as possible will not serve that purpose.

Insight into the understanding of the calculation procedure can be gained by having:

- Students explain orally or in writing the thinking that is part of a computational procedure.
- Students make selections from given choices of responses and then defend the choices by explaining, orally or in writing, why the alternatives were not chosen.

Possible Types of Assessment Tasks

1) **6 x 12 = ☐   8 x 235 = ☐   16 x 24 = ☐**

   a) *Who do you think would want to calculate the answers for these equations? Try to make up a word problem for the equations.*

   b) *Explain how you would estimate the answer for each equation. Do you think the answer is greater than or less than your estimate? Explain your thinking.*

   c) *How would you explain the steps for calculating the answers for these equations to a younger student? What are the important ideas that younger students should learn and remember? Why are they important?*

2) *What mistakes do you think were made in each of the calculations? Why do you think the mistakes were made?*

```
  25        634        32
 x 8        x 7      x 19
 160        420        18
            210       270
             28         2
            658        30
                      320
```

   What are the correct answers?

3) *Show two different ways of calculating the answer for* **24 x 12 = ☐**.

4) *Explain how a calculator can be used to calculate the answer for* **24 x 12 = ☐** *in two different ways.*

Reporting

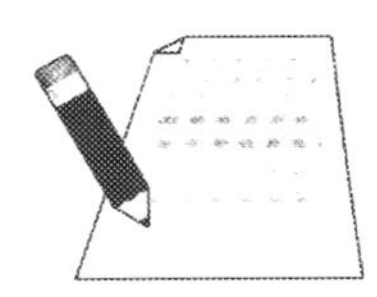

The suggested types of assessment tasks included collecting data about equations of the following type: **8 x 32 = □**; **7 x 649 = □**; **46 x 51 = □**.
The assessment data that are collected can make it possible to share information about:

- Indicators of ability to *connect* based on answers to the question about who would want to solve these types of multiplication equations and the word problems that were generated.

- Indicators of *number sense* based on the *estimation* strategies that were employed and the ability to explain whether or not an estimate is greater than or less than the answer.

- Indicators about ability to *visualize* numbers based on making reference to the meanings of digits with respect to their place value and knowing the meaning of the partial products that have to be calculated as part of the computational procedure.

- Indicators of *flexible thinking* about numbers based on calculating the product for an equation in more than one way.

- Indicators about understanding the computational procedure or algorithm based on the corrections that were suggested for mistakes that were shown and being able to explain the reasons for each step that is followed as part of a calculation procedure.

For Reflection

1) What possible rationale might there be for increasing the number of digits in both factors as part of the teaching of the multiplication algorithm? What would be part of an argument against this type of extension?

2) What would you say to people who present the conclusion that as far as they are concerned the skill level of students in the mathematics classroom is deteriorating and is not what it was when they went to school?

3) What would you say to someone who claims that the main reason for the inability of students to be able to *do mathematics well and quickly in their heads* is the presence and use of the calculator?

4) Think of a page of twenty multiplication equations with two digit factors along with the request *Calculate the Products*. What possible goals for students might an author of such a page have in mind? What possible new skills or ideas can be learned by completing a page of this type?

5) What understandings and generalizations about multiplication would enable a student to calculate the answer for **12 345 x 56 789 = □** on a calculator and without a calculator?

Division Beyond the Basic Division Facts

After a presentation one high school teacher shared the observation that not one of her students had an understanding of the division algorithm.[7] When she was asked why she thought that was the case, she attributed it mainly to the symbol $\overline{)\quad}$ used by the students for long division.

Transcripts from many diagnostic interviews with students as well as conversations with adults support the teacher's observations. The computational procedure for long division is not understood. What are some possible reasons for this being the case?

First and foremost is the fact that the presentation of the computational procedure was and is rule-based in most classrooms. In some classrooms the steps to be followed are displayed on the wall and the procedure to calculate an answer becomes a ritual of following a sequence of the listed or memorized steps.

Another contributing factor for the lack of understanding is that the language used while calculations are made is inappropriate. The terminology and the recitals for the steps to be followed do not contribute to the *visualization* of the divisive action nor the numbers behind the numerals that are manipulated.

Sample responses from diagnostic interviews illustrate some of the confusion inappropriate use of language can lead to.
The majority of students will read and interpret **6 ÷ 2 = □** as, *six divided by two* or as, *two into six*. When asked whether they know another way of reading it, there is no hesitation for many. They will state, *two divided by six* or, *two into six* and here is evidence that the symbol $2\overline{)6}$ and reading the statement from left to right using the slang *goes into* or *into* are more than likely responsible for leading many students to believe that it can be read in either way.

It is a somewhat sobering experience to encounter students who believe that division equations can be read in either way. For example:

- **5 ÷ 9 = □** becomes, *'five divided by nine – it will go once with a remainder of four.'*
- **0 ÷ 2 = □** and **6 ÷ 0 = □** fall into the same category – *'anything divided by zero goes zero times.'*

A collection of expressions for the same equation, **8 ÷ 2 = □**, used not only by subjects during interviews but by different teachers as part of their lessons shows how confusing the division procedure can become for students:

- *eight divided by two*
- *two goes into eight*
- *eight into two parts*
- *eight into two groups*
- *two into eight*
- *eight into groups of two*
- *eight take away groups of two*
- *one-half of eight.*

The last expression leads to all kinds of difficulties. It probably is responsible for the confusion that exists when an attempt is made to interpret $\mathbf{8 \div \frac{1}{2}}$. During interviews subjects have interpreted it as: *'one-half of eight'*; *'eight divided by two'* or *'eight into two parts.'*

Attempts by students or adults to explain the steps that are taken to calculate an answer for a task that involves long division indicate rote procedural knowledge. Reasons for the moves that are made may not be known. The digits that are part of the calculation are referred to as ones or as tens and ones. Rarely is someone able to explain the meanings of the partial quotients and partial products.

One attempt to begin to get rid of some of the confusion and lack of understanding that exists would involve being consistent with reading division equations from left to right when the symbol ÷ is used and reading from the 'inside to the outside' when $\overline{)\quad}$ is used. This consistency would serve students well since it would get rid of *goes into* or '*Guzinta*'[8] (p.140).

> In school I have an awful time to learn 'rithmetic.
> That's the one thing that gets my goat, and the teacher thinks I'm sick;
> She makes me learn guzinta, 'though I declare I won't;
> There are some things I can understand – but other things I don't.
> Guzinta should be easy work for kids as smart as me.
> Why, I know in an automobile the gear guzinta three;
> And when I've finished my lessons, you'll have a chance to see
> How much bread and jam guzinta a hungry kid like me!
> Five guzinta ten and ten guzinta twenty.
> And when I don't know my 'rithmetic, teacher goes into plenty!
> But five what guzinta ten what, I never can see –
> 'N that's why no 'rithmetic ever guzinta me!
>
> Author unknown

When the topic of long division comes up during discussions with adults, many are not reluctant to state how much they hated it. Some classify it as being difficult mathematics. This conclusion might be related to the fact that some teachers made reference to *hard math* or *now comes the difficult math* when the computation procedure was introduced. One teacher who was in charge of a group of special needs students shared the observation that whenever he mentioned long division or fractions, many of his students expressed an urge to visit the washroom.

Personal Computation Strategies

It is unlikely that any student will develop a recording procedure for a personal computational strategy involving division that in any way comes close to the standard algorithm that is used in the schools. However, it is possible to introduce the recording procedure for the division algorithm without having to present any new major skills or ideas. If students have the necessary prerequisites, they can be shown how to use the ideas they know and record results of calculations in a new format.

Pre-requisites for Personal Computation Strategies

The pre-requisite skills and understandings that will enable students to understand all parts of the calculations they perform and know the meanings of the digits they are manipulating and recording include the following:

- Both of the interpretations of division, the *equal grouping* and *equal sharing* need to be known. These interpretations are part of every day experiences which means that both are required for solving problems that relate to these experiences. The *equal grouping* interpretation makes it possible to derive the basic division facts should there be a need for it. The *equal sharing* interpretation aids in the ability to *visualize* the steps that are part of the computational procedure beyond the basic facts.

- The *basic division facts* need to be known which means knowing that for most *basic division facts* there exist matching *basic multiplication facts*.
  For example,
  63 ÷ 7 can be solved by applying the *equal grouping* interpretation and thinking, *how many groups of seven are equal to sixty three*?
  Knowing that *basic division facts* can be derived from matching *basic multiplication facts* enables students to get unstuck or to re-invent *basic division facts* that may have been forgotten.

- *Number sense*, or the ability to *visualize* numbers, will enable students to know the meaning of the partial quotients and partial products that are part of the computational procedure. *Flexible thinking* about numbers will enable students to *visualize* a number in a different way and assign a different name to it.

Personal Computation Strategies

The introduction to a computational procedure can start with a meaningful problem that needs to be solved. For example,

Seventy two students are to be assigned to three teams for the annual treasure hunt. How many students will there be on each team?

The students are requested to work with a partner, to show the number seventy two with base ten blocks and then decide how they would assign these blocks or people to three groups so that each team has the same number of competitors.
The same setting could involve the request of having to decide how to share **$72** equally with three people.

After solution strategies are explained and compared, the task of how to record the thinking that was part of these procedures is presented. A reminder may be needed about the two possible ways of interpreting and reading the symbol $\overline{)\ \ }$, as *divided by* and as *shared with*.
$\mathbf{3\overline{)72}}$ is read as, *seventy two divided by three* or, *seventy two shared with three people or with three teams or groups*.

The seven tens are shared equally with the three people. That means each person gets two tens and one ten remains to be divided. Three possible ways to record the thinking up to this point include:

a) $\begin{array}{r} 2\ \text{tens} \\ 3\overline{)72} \\ -\ \underline{6\ \text{tens}} \\ 1\ \text{ten} \end{array}$ b) $\begin{array}{r} 20 \\ 3\overline{)72} \\ -\ \underline{60} \\ 10 \end{array}$ c) $\begin{array}{r} 2 \\ 3\overline{)72} \\ -\ \underline{6} \\ 1 \end{array}$

When these three recording procedures are compared during a discussion, the request is made to point out possible advantages or disadvantages that might be noticed.

Any recording procedure that keeps reminding students of the meaning of the numbers behind the numerals that are used and recorded is advantageous since it can contribute to *visualization*.

Even if the third recording procedure is adopted, the actual value of the digits should be referred to while performing calculations; that is two tens and six tens are recorded and not two and six. Making reference to the actual meanings reinforces the reason for recording the two tens above the seven tens in the dividend.

The one ten and the two ones, or twelve dollars now need to be shared. After four dollars are given to each person, all of the money is taken care of.

For the first two recording procedures, the second partial quotient is recorded above the first partial quotient and then the two partial quotients are added to give the answer.

```
       24                   24
        4                    4
        2 tens              20                   24
a)  3)72           b)   3)72          c)    3)72
      - 6 tens            - 60               - 6
        12                  12                 12
      - 12                - 12               - 12
         0                   0                  0
```

When these three recording procedures are compared, the focus should be on how they are similar, rather than focusing on how they differ. The same numerals are part of each recording procedure. As numerals are pointed to, requests are made to state why they are recorded in their respective positions.

After two or three examples are solved in three different ways, the request is made to explain which of the three recording methods is preferred and why.

Conceptual understanding of division and *number sense* will make it possible to transfer the recording procedures for two-digit dividends to three-digit dividends and beyond. As the following equations are presented to pairs of students, they are invited to calculate the answers by thinking about the *equal sharing* action:

632 ÷ 4 = ☐ **7 425 ÷ 3 = ☐**.

The suggestion could be made to think of sharing **$632** and **$7 425** with four and three people, respectively.

The guidance that is provided could include:

- *Describe how the money could be shared equally.*
- *Be ready to explain your strategy with base ten blocks.*
- *Try to think of a way of recording the steps you followed to get the answer.*

After different strategies and recording procedures have been solicited and described comparisons are made.

- *What is the same about the strategies and the recordings, and what is different?*
- *How would you explain the steps of your calculations to a younger student?*
- *What are the important ideas that a younger student should learn? Why are they important?*

Challenge – a Two-digit Divisor

Conceptual understanding of division and *number sense* will enable many students to be ready for the challenge of transferring what they have learned about one-digit divisors to a calculation procedure that involves a two-digit divisor.

To give students an opportunity to develop a personal strategy, they could be asked to try and think of how they might calculate the answer for an equation like

384 ÷ 12 = □.

The following types of comments could provide some guidance to the students:

- *Think about how twelve people would share* **$384** *equally.*
- *Be ready to act out your strategy with base ten blocks.*
- *Think of a way of recording on paper how you calculated the answer.*

Interpreting Remainders

Students need to learn that dealing with remainders depends on the type of problem that is solved. Two possible ways of reaching this conclusion include the following settings.

- A simple equation and the different methods of recording or interpreting the remainders are presented. The challenge is presented to try and make up a matching word problem for each scenario.
  For example, **25 ÷ 2 = □**
  Possible answers: a) **13** b) **12 rm 1** or **12** c) **12 1/2 or 12.5**
  Do the problems that are created illustrate in a meaningful way:
  - when an answer is rounded up.
  - when there is a remainder and it is ignored or accommodated in a special way.
  - when the remainder can be thought of as a fraction.

- A second possibility consists of listing the possible answers for an equation and presenting four word problems. The request is made to match the problems to the answers and to explain the thinking.
  For example, for **25 ÷ 2 = □** the following problems could be presented:
  - *Two children equally share twenty five pennies. How many will each child get?*
  - *Twenty five children have to be assigned to two teams that have the same number of players. How many players on each team?*
  - *Two persons share twenty-four small pieces of fudge. How many pieces of fudge could each person get?*
  - *Twenty five horses have to be moved to a different pasture. A trailer that is used can hold two horses. How many trips will it take to move all the horses?*

Accommodating Responses

Different students or groups of students may come up with solutions that are unique and differ from the choices that were given. What do students suggest should or could be done with the extra penny or the extra player? Why are different answers possible for the question about the pieces of fudge?

Appropriate Practice

The goal of appropriate practice settings is to present exercises that give insight into the calculation procedure and involve *mathematical reasoning* or *analysis*.

Types of Activities and Problems

- **Missing Numerals – Meanings of Partial Quotients, Products and Remainders**
  The request is made to supply the missing numerals in computational procedures and to explain orally or in writing how decisions were made. Questions are posed about partial quotients and partial products.

```
   1□2 rm□
5)578
  3
  27
  2□
   0□
    6
    □
```

  Possible types of questions:
  - *Why is the* **1** *recorded above the* **5**?
  - *What does the* **3** *below the* **5** *mean?*
  - *What does the* **27** *mean?*
  - *Why is it impossible to know what should be done with the remainder?*

- **Arranging Digits According to Instructions**
  A division equation is presented along with a sufficient number of digits for the dividend and the divisor. These digits could be randomly drawn from an envelope containing the numerals **1** to **9** or several students could be asked to state their favourite numerals between **0** and **10**.

  One of the following equations is presented.

  a) □□□ ÷ □ = □ **2, 3, 4, 5**

  b) □ □ □ □ ÷ □ = □ **1, 2, 3, 4, 5**

  Requests are made about the placement of the digits.
  For example,

  Determine and record the division equation that has:
  - *the greatest answer.*
  - *the least answer.*
  - *an answer between the greatest and the least answer.*
  - *the greatest remainder.*
  - *the least remainder.*

  As pairs of students attempt these tasks, the request could be made to keep track of the decisions that are made and the different numerals that are tried. The responses to the request are compared during a follow-up discussion.

- **Finding the Mistakes**

  Computational procedures that include errors are presented. The request is made to identify and explain the errors and to make a guess about why these errors might have been made.

  The examples of errors can include:
  - The zero is omitted in the partial product in the tens place.
  - The remainder is recorded as part of the quotient.
  - The remainder that is recorded is equal to or greater than the divisor.
  - The partial product is recorded in the incorrect place value position.
  - The partial dividend is recorded in the incorrect place value position.
  - The computation includes an incorrect multiplication fact.
  - The calculation includes an incorrect subtraction fact.

- **Estimating**

  For one type of task, equations along with two choices of estimates arepresented. The following requests are made,
  - *Without the use of pencil and paper decide which of the two choices do you think is a closer estimate to the answer.*
  - *Be ready to explain how the decision was made.*
  - *Do you think this estimate is greater than or less than the answer? Explain your thinking.*

  For example:

  146 ÷ 4 = ☐ *Is the answer closer to* **30** *or to* **40**?
  260 ÷ 8 = ☐ *Is the answer closer to* **30** *or to* **40**?
  5 319 ÷ 9 = ☐ *Is the answer closer to* **500** *or to* **600**?
  3 165 ÷ 8 = ☐ *Is the answer closer to* **300** *or to* **400**?
  2 985 ÷ 7 = ☐ *Is the answer closer to* **400** *or to* **500**?

  After an equation like **1 408 ÷ 6 =** ☐ is presented the request is made to estimate the answer. As part of the discussion, students are requested to:
  - *Describe the estimation strategy that was used.*
  - *Tell whether the estimate is greater than or less than the quotient.*

- **Missing Parts of an Equation**

  The problem of identifying missing numerals in division equations is presented.
  - *What do you think is missing?*
  - *Describe how you try to determine what is missing.*
  - *How do you know you are correct?*

  732 ÷ ☐ = **122** rm **0**
  815 ÷ ☐ = **116** rm **3**
  7 520 ÷ ☐ = **940** rm **0**

Assessment Suggestions

A student in grade five who was recommended for a diagnostic interview brought with him several activity sheets. One of these consisted of twenty two similar division equations. The student had attempted seven. Four of these were correct and two were partially correct in that the computational procedure was followed and some of the partial quotients and products were correctly calculated. The items that were not attempted all had a red x mark beside them and the score on the sheet was **4** out of **22**.

This student lacked self confidence. The fact that items that were not attempted had been marked as incorrect may have contributed to this state. The results of the interview indicated that the main aspects of mathematics learning that required intervention related to mental mathematics strategies of the basic multiplication facts and to ability to visualize numbers. More practice with items of the same type would not have helped this student.

The reason for including this anecdote is related to the question about how to assess an understanding of the computational procedure for division. The very low score that was assigned to this student would change dramatically if only the attempted items were considered. One thing that is certain is that the score that was assigned to this student does not tell anything about how well aspects of the computational procedure are understood and this would also be true for a perfect score on this timed test.

Possible Types of Assessment Tasks

1) A partially completed item is presented and students are asked to answer questions and supply missing numerals. For example,

```
      9□0 rm2
  8)7 522
   - 72
      32
    - □2
      □□
```

The types of questions can include:

- *Why isn't there a numeral recorded above the* **7**?
- *Why is the* **9** *recorded above the* **5**?
- *What is the meaning of the* **72**? *How was it calculated?*
- *What is the meaning of the* **32**? *How was it calculated?*
- *What numeral should be recorded above the* **2** *in the tens place and why should that numeral be recorded there?*
- *Why does a* **0** *appear above the* **2** *in the ones place?*
- *When an equation has a divisor of* **8**, *what is the greatest possible remainder and why is it the greatest possible remainder?*
- *How do you know or how could you find out that the missing numerals that were filled in are correct?*

2) The request is made to describe, orally or in writing, the strategy used to estimate a quotient and the thinking for the decision whether or not the estimate is greater than or less than the quotient.

For example,

- *How would you estimate the answer for* **3 894 ÷ 6 = □**?
- *Will the answer be greater than or less than your estimate? How do you know?*

continued next page

3) The identificationn of errors and describing possible reasons for these can give insight into understanding.

a) *What mistakes do you think were made?*
Why do you think the mistakes were made?
For example,

```
      53 rm4
  3)1 513
    15
      13
       9
       4
```

b) The request is made to observe a simulation to determine a quotient for an equation with base ten blocks or with play money that includes incorrect 'moves' or incorrect place value representations. The students are invited to describe, orally or in writing, reactions to the simulation.

Reporting

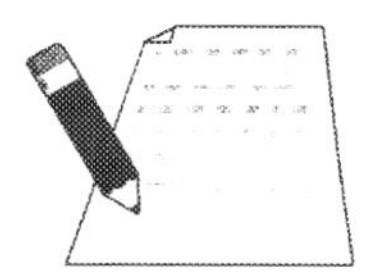

The types of assessment tasks that are suggested can make it possible to share information about indicators of:

- *Conceptual understanding of division*: A simulation of the divisive action and talking about sharing as the computational procedure is explained yields information about this understanding as well as the ability to *connect.*

- *Number Sense*: Recording partial quotients in appropriate place value positions and knowing the meanings of the partial products imply that numbers are *visualized.*

- Ability to *estimate*: The ability to estimate and being able to tell whether or not an estimate is greater than or less than the answer are also indicators of *number sense.*

- Understanding the computational procedure: Data about this understanding can be collected as students correct errors in answers, recording procedures or use of materials and explain why they think these errors were made. Ability to explain different interpretations of remainders is another possible indicator of under-standing.

For Reflection

1) Assume that the teacher of the grade five student who received **4** out of **22** on his activity sheet of solving division equations believes that it is important to give students timed activities and timed tests for computational procedures. What questions would you ask of that teacher? Why would you ask these questions?

2) Assume that the teacher of the student in grade five believes that tests and activity sheets should be collected once one-fourth of the students have completed all of the items. What questions would you ask of this teacher? Why would you ask these questions? What do you think this teacher should say to parents whose children get a score of **4** out of **22** on these tests?

3) What would you say to people who claim that increasing the level of difficulty for division by increasing the number of digits in the dividends and divisors is good for the brain; it *toughens up the muscles of the mind*?

Chapter 5 – Computation Procedures – Personal Strategies

Decimals (+, –); Fractions (+, –); Percent Problems; Integers (+, –)

The statement that *number sense* is the key to *numeracy* can be illustrated and reinforced with discussions about the operations with decimals. Without *number sense* the focus of teaching about the operations with decimals has to be on direct teaching and practice. *Number sense* and *conceptual understanding* of the operations enable students to develop personal strategies.

Addition and Subtraction with Decimals

Pre-requisites – Key Ideas from the Previous Grades

The types of activities that were part of previous mathematics learning and the problems that were solved in these settings include the following pre-requisite skills, procedures and ideas:

- The students have conceptual understanding of addition and subtraction.
- The students know the basic addition facts and basic subtraction facts.
- The students have begun to develop *decimal number sense*.
  That means that they can:
  - *Visualize* decimals. They can represent decimals with base ten blocks and they can think about them in terms of distances and points on a number line.
  - *Think flexibly* about decimals. They know that a decimal can be represented with base ten blocks in different ways and can be named in different ways.
  - *Estimate* decimals. They know whether a decimal less than one whole is closer to zero, to one-half or to one whole.
  - *Relate* decimals. They can compare two or more decimals and put three or more decimals in order from least to greatest or vice versa.
- The students have developed personal strategies for the addition and subtraction of whole numbers. These strategies include the main generalization that it is important to keep track of digits that are in the same place value positions.

Addition of Decimals

The presence of *decimal number sense* makes it possible to reach important learning outcomes without having to follow a hierarchical instructional sequence that begins with tasks for tenths and then moves on to hundredths and thousandths. Decimal *number sense* and *conceptual understanding* make it possible to use a teaching *through problem solving* setting which enables students to develop *personal strategies* for computations and derive important generalizations. The example that is described illustrates possible components of a teaching *through problem solving* setting:

A problem and a set of questions are presented. The students work alone or with a partner to come up with responses for the requests.

There are four pieces of string of different lengths:
0.5 m **4** m **0.075** m **2.21** m
If all of the pieces of string were placed end to end what would the total length be?

The following types of requests can be made:

- *Before doing any calculations, what is your estimate for the total length? Be ready to explain your strategy.*
- *Do you think your estimate is less than or greater than the total length? Explain your thinking.*
- *Add the lengths to determine the total length. How is adding the decimals the same as adding names for whole numbers?*
- *Is the addition of decimals in any way different from the addition of whole numbers? Explain your thinking.*
- *If you were to make up a rule for adding decimals what would it be?*
- *If you were to teach younger students how to add decimals what would you tell them that you think they should know?*

Rather than using pieces of string, the decimals can be represented with base ten blocks. These blocks are placed to make them easily visible for the students. After each set of blocks is labelled **0.5**, **4.0**, **0.075** and **2.21,** respectively the questions are changed to make reference to calculating the sum.

Accommodating Responses

Experiences in classrooms and responses elicited during diagnostic interviews indicate that the rules that some students generate for the addition of decimals require redirection. If decimals like those in the example are used, many students' rules will make reference to, '*you can add zeros*' or, '*you can add as many zeros as you need or want.*'

Comments like these provide an excellent opportunity to show how important it is to use correct language. One example is all it takes to convince most of these students of this fact. When students are shown what happens when these suggestions are followed, i.e.,

2.21 + 0 = □ **0.5 + 0 + 0 = □**

many students will quickly indicate that something different was meant and they will correct themselves.

Appropriate Practice

Practice settings should remind students of the importance of keeping track of digits that have the same meaning. These settings should reinforce the importance and power of estimating.

Types of Activities and Problems

Placing the Decimal Point
Equations are presented that have missing decimal points:

- *in the sum.*
- *in the addends.*
- *in the addends and the sum.*

For example:

3.12 + 0.99 = 411
44 + 24 = 4.64
44 + 24 = 6.8
44 + 24 = 0.68
44 + 24 = 2.84
44 + 0.24 = 4424

Where do you think the decimal points belong to show equations are true?
Try to make decisions without the use of pencil and paper calculations.
Be ready to explain your strategies.

Finding and Correcting Mistakes
Examples that include mistakes are presented.

What are the mistakes
Why do you think they were made?

The examples of errors could include:

- The decimal point is not lined up for all of the addends.
- The decimal point is omitted in an addend and in the sum.
- The second of two addends is annexed to first addend to determine the sum.
- The whole number part is omitted from the addition of two or more mixed decimals.

Selecting Addends
Several equations with missing addends and a list of decimals are displayed.
For example:

□ + □ = 2.64 　 □ + □ = 26.42 　 □ + □ = 0.624 　 □ + □ = 2.640

24.02 　 0.24 　 24.24 　 0.024 　 0.42 　 2.4 　 0.204

The following requests are made:

Select two decimals for each equation that you think will result in an equation that is true.
Try to make your decision without pencil and paper calculations.
Be ready to explain the strategies you used to make decisions.

As a variation to this setting equations with three addends are shown. These equations could have one, two or three missing addends.

Missing Numerals
Equations with missing numerals are presented.
For example,

2.0□ + □.08 = 10.□9 8.□□ + □.2□ = 12.69

What do you think the missing numerals are or what could they be?
Be ready to explain the strategy that was used.

Subtraction of Decimals

An introductory setting can include a problem setting of the following type:

- The requests are made to make up a word problem for an equation, i.e., **0.68 – 0.37 = □**, to calculate the answer and to come up with a statement that compares subtraction of decimals to the subtraction of names for whole numbers. What is the same? What is different?
- The decimal **0.2** is shown with base ten blocks. Pairs of students are challenged to illustrate how calculating the answer for **0.2 – 0.15 = □** can be shown with the base ten blocks and to record on paper what was done with the blocks?
- The request is made to calculate the answer for **3.1 – 0.55 = □** and to be ready to show how to illustrate the solution procedure with base ten blocks.
- The request is made to make up a rule for subtracting decimals. The rules are compared. Rather than making up a rule, the request could be to come up with an explanation that would tell a young students how to subtract decimals.

Appropriate Practice

The types of activities and problems can be similar to those that were described for the addition of decimals.

Types of Activities and Problems

- **Placing the Decimal Point.**
  Where do you think the missing decimal points belong to show equations that are true?
  Try to make decisions without the use of pencil and paper calculations.
  Be ready to explain your strategy.

68 – 49 = 1.9 **68 – 49 = 0.19** **68 – 49 = 67.51**
157 – 3.1 = 12.6 **1.57 – 31 = 1.26** **157 – 0.31 = 126**

- **Finding and Correcting Mistakes.**
  The types of mistakes that are presented could include:
  - Omission of a decimal point.
  - Incorrect placement of the decimal point in a difference.
  - A numeral is subtracted from a numeral in an incorrect place value position.

- **Selecting the Minuends and the Subtrahends for Given Differences.**
  The Several equations with missing minuends and subtrahends are presented. The request is made to select from a list the missing decimals to record equations that are true.
  For example:

  □ – □ = 3.17   □ – □ = 31.7   □ – □ = 0.317

  0.086   0.86   8.6   86.0   40.3   4.03   0.403

  Select and record the decimals that you think will show equations that are true.
  Be ready to explain your thinking.

- **Placing Decimal Points in Minuends and Subtrahends to Show Correct Differences.**
  Place decimal points to show true equations.
  Be ready to explain your thinking.

  220 – 99 = 1.21   220 – 99 = 12.1   220 – 99 = 0.121   220 – 99 = 2.101

- **Missing Numerals.**
  Equations with missing numerals are presented.
  For example,

  □6.4 – 9.□ = 26.6   4.□3 – □.86 = 3.1□   □.4□ – 1.3□ = 6.04

  What do you think the missing numerals are or what could they be?
  Be ready to explain the strategy you used.

Assessment Suggestions

To gain insight into students' understanding, opportunities can be given to have them explain orally or in writing, what they are thinking as they are solving an equation. Another option could be a multiple choice format that requires students to defend the choice they make and state reasons for not selecting the alternatives.

Possible Types of Assessment Tasks

- Listing the important things that need to be remembered when decimals are added. Why are they important?
- Listing the important things that need to be remembered when decimals are subtracted. Why are they important?
- Explaining how the addition and subtraction of decimals is different and how it is the same from adding and subtracting other numerals.
- Correcting errors in decimal point placements in a sum and in a difference and explaining the possible reason or reasons for the errors.
- Placing the decimal points to show the correct sum and the correct difference. Explain why the placement is correct.
- Selecting from several choices the correct sum and the correct difference for given equations. Explain why the choices are correct and why the alternatives are incorrect.
- Placing decimal points to create equations that are true and explaining the strategy that was used.
  For example,

  34 + 42 = 4.54   **61 – 29 = 5.81**

Reporting

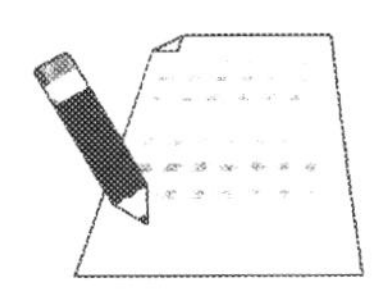

Reports could inform readers about the meaning of understanding how to add and how to subtract decimals. Indicators of this understanding that have been observed and noted can be shared.

The ability to estimate enables students to make predictions about answers and to check on the reasonableness of answers. This ability also tells students whether or not decimal points appear in the correct places or where to place decimal points in order to show correct answers or to record true equations. Since this is the case, reports should inform readers about possible indicators of ability to estimate that have been observed and collected.

The activities and problems that were part of teaching about addition and subtraction of decimals and the suggested types of assessment tasks may make it possible to note and record indicators of:

- Ability to *visualize* decimals.
- *Flexible thinking* about decimals.
- *Confidence and willingness to take risks.*
- Ability to communicate mathematically.

Any of these indicators could be included in reports.

Multiplication and Division with Decimals

The learning outcomes in the curriculum identify the introduction of one digit whole number multipliers and one digit whole number divisors for decimals. Decimal *number sense* and *conceptual understanding* of multiplication and division should enable students to reach these goals without any difficulty and even enable them to go beyond the outcomes.

Pre-requisites – Key Ideas from the Previous Grades

The key pre-requisites that will make it possible to develop *personal strategies* for the multiplication and division of decimals include:

- An understanding of the *groups of* interpretation of multiplication.
- An understanding of the *equal sharing* interpretation of division.
- Important aspects of decimal *number sense* which include the ability to: *visualize* decimals; *think flexibly* about decimals; *relate* decimals and *estimate decimals*.
- *Personal strategies* for the multiplication of whole numbers.
- An understanding of a computational procedure for the division of whole numbers by a one digit divisor.

Multiplication with Decimals

What would the main learning outcomes for multiplying tenths, hundredths and thousandths by a single digit whole number multiplier be for students who have the required pre-requisites? What generalization about this multiplication will enhance and be most useful to future learning of any type of multiplication with decimals?

Conversations with adults and responses collected during diagnostic interviews indicate that at one time the major generalization for the multiplication of decimals related to counting the number of decimal places in each factor and then ensuring that this number appeared in the product. When many adults and students are shown:

15.5 x 8.24 = 12 772 (as a calculator would display the product)

and are then asked to place the decimal point to indicate the answer, they will count three decimal places from the right because there are three decimal places in the factors. Often this placement of the decimal point is supported with a comment like,

'That is the rule we were taught.'

After one grade seven student had placed the decimal point between the **2** and the **7** he stated, [2] *'...because eight point twenty four has two decimal places in the number and fifteen point five has one place ... though ... eight times fifteen is one hundred twenty ... so it would go there ...(**127.22**) – Oh ...Yeah!'* (the dots indicate pauses).

When some people watch this student's comments on a video, they actually pose the question,

'So why did they teach us the rule?'

The adjustment made by the student during the interview reinforces the importance of *number sense*. It also illustrates the key generalization related to the importance of estimating the products. Any rule about the counting of the number of decimal places in the factors has to include the proviso that before the decimal point is placed in the product, estimation needs to be used to confirm its placement.

One possible introductory setting to the multiplication with decimals could include the presentation of problems along with choices for products and requests related to these choices. For example,

There are **6** boxes. Each box weighs **0.8** kg.
What is the total weight of the boxes? **48** kg **4.8** kg **0.48** kg

There are **4** stacks of coins. Each stack has **$0.24**.
What is the total amount of money? **$96.00** **$9.60** **$0.96**

There are **2** strips of paper. Each strip is **0.252** m long.
What is the total length of the strips? **5.04** m **0.504** m **50.4** m

The following requests are made as the examples are presented, one at a time:

Select the choice of an answer that you think is obviously wrong?
Why do you think so?
Is there another choice that you think could not be the answer?
Why is that the case?
Which do you think is the answer? Why do you think so?

After the correct answer for each example has been selected, the request is made to carry out the computation to show the answer.

Look at your calculations. How is multiplication with a decimal different from multiplying two other numerals and how is it the same?
What do you think young students should know about multiplying with decimals to calculate answers?

The goal is to reach conclusions that include ideas about:

- The importance of estimating the product.
  This estimation may begin with being able to tell whether or not a product is close to one whole, close to one-half or close to zero.
  The estimation could focus on possible choices that are obviously incorrect and explaining why that is thought to be the case.
- The correct placement of the decimal point in products and explaining the strategies that were used to make decisions.
- Carrying out the computation for multiplication as they did for whole numbers and knowing where to place the decimal point.

One type of activity that can contribute to reinforcing the importance of estimation involves the presentation of equations with factors where the first partial product results in a zero for the lowest place value position, a zero that is not displayed as part of a product on a calculator. The task consists of placing the decimal to show the correct product and to be able to explain the estimation strategy used to make decisions.
For example,

8 x 0.9 = 72 **4.72 x 5 = 236** **1.25 x 4 = 5** **0.401 x 7 = 2807**
0.222 x 5 = 111 **6 x 0.5 = 3** **0.006 x 2 = 12** **5.69 x 6 = 3414**

Appropriate Practice

The main goal of practice tasks is to provide more opportunities to employ and refine *estimation strategies* and to use aspects of *mathematical reasoning*. The types of tasks that can be presented are similar to those for addition and subtraction of decimals and to those that were part of the teaching sequence

Types of Activities and Problems

Placing the Decimal Point

Two types of tasks can include placing the decimal point in the digits to show the correct product or placing the decimal point in the factors to make the shown product the correct answer. For either setting, the request is made to explain the estimation strategy used as part of the decision making. As part of the tasks, an example that has more than one possible correct response could be included. If deemed appropriate an example that is not possible could be included.
For example,

Where does the decimal point go to show correct answers?
Explain your thinking.

3.72 x 2 = 744 **0.42 x 8 = 336** **0.506 x 5 = 253**
2.7 x 5 = 135 **1.005 x 6 = 63**

Where does the decimal point go to make the shown answers correct?
Explain your thinking.

64 x 5 = 32.0 **36 x 7 = 2.52** **4 x 3 = 0.12**
5 x 1 = 0.005 **1 805 x 3 = 5 415**

During the follow-up discussions the strategies that were used are compared and calculations can be shown on the chalkboard to indicate that the choices that were made are correct, that more than one choice is correct, or that a true equation is not possible.

Finding and Correcting Mistakes

The different types of errors or combinations thereof could include:

- Basic multiplication fact error.
- The decimal point is omitted in the product.
- The decimal point is placed incorrectly in the product.

For example,

What do you think is wrong? Why is it wrong?

0. 09 x 8 = 7.2 **4.72 x 5 = 236** **6 x 0.5 = 0.3** **0.8 x 0 = 0.8**

Selecting Factors

Products are shown for multiplication equations. The task consists of selecting from a list of choices two factors that would yield that product.

Select the two decimals that you think will give the answer that is shown? Explain your thinking.

□ x □ = 0.024
□ x □ = 2.4 **3.0 0.3 0.03 0.003 8.0 0.8 0.08 0.008**
□ x □ = 0.24

Division with Decimals

The main learning outcomes for dividing decimals to thousandths by a whole number include the ability to:

- Look at an equation and tell at a glance whether or not the quotient will be less than one whole, about one whole or greater than one whole.
- Estimate a quotient, tell whether or not the estimate is greater than or less than the quotient and to explain the reasoning for the decision.
- Use estimation to place the decimal point in given digits to show the quotient for an equation and to explain the thinking.
- Use base ten blocks and the *equal sharing* interpretation to simulate the divisive action to calculate quotients.

When the divisor is a whole number, the *equal sharing* interpretation enhances the ability to *visualize* the action that is part of each step of the computation procedure.

Problems about equal sharing could be presented. Before the choices of quotients for each problem are presented the request is made to respond to,

Do you think the answer will be greater than one, about one or less than one? How do you know?

After the choices for the answer are displayed, the students are asked,

Which choices do you think are obviously not the answer? Why is that the case?

Sample problems:

Four people want to share a fish weighing **3.6** kg.

3.6 ÷ 4 = □

How much fish will each person get? **9 kg** **0.9 kg** **0.09 kg**

Three children are asked to share **$1.26**.

$1.26 ÷ 3 = □

How much money will each child get? **$42.00** **$4.20** **$0.42**

4.684 m of ribbon is to be shared between two bulletin boards.

4.684 ÷ 2 = □

How much ribbon will there be for each? **234.20 m** **23.42 m** **2.342 m**

As was the case for multiplication, the request is made to use the familiar computation procedure for division and to show how the answers could be calculated. The request is made to be ready to explain and show how calculating the answer could be illustrated with base ten blocks.

The following challenge could be provided,

If you were to teach a young student about division with decimals,
what important things do you think you should include in your teaching?
Why is that the case?
How would the young students know where to place the decimal point in the answer?

Appropriate Practice

The goal of activities with division of decimals by a whole number is to reach conclusions that include the following ideas:

- If the dividend is greater than the divisor, the answer will be greater than one.
- If the dividend is less than the divisor, the answer will be less than one.
- If the dividend is close to the divisor, the answer will be close to one.
- One strategy of estimating the quotient begins by looking at the whole number part of the decimal. If a decimal is less than one, the digit in the highest place value position is used as part of an estimation strategy.
- The calculation procedure is the same as for numbers that are not decimals and the decimal point is placed in the answer when there are no more numbers greater than one to be shared.

Types of Activities and Problems

- **Placing the Decimal Point**
  The request is made to place the decimal point to show the answer and to be ready to explain the thinking.
  For example,

  Where does the decimal point belong to show the correct answers?
  How do you know?

  0.88 ÷ 8 = 11 **6.6 ÷ 6 = 11** **2.05 ÷ 5 = 41**
  4.92 ÷ 6 = 82 **0.61 ÷ 2 = 305**

Where does the decimal point belong to make the shown answers correct? How do you know?

72 ÷ 8 = 0.9 **136 ÷ 4 = 0.34** **308 ÷ 7 = 4.4**
264 ÷ 8 = 3.40 **186 ÷ 6 = 0.031**

During the follow-up discussions the strategies that were used to make decisions are compared and calculations can be shown on the chalkboard to show that the decimal placement was correct.

- **Finding and Correcting Mistakes**
  Examples can be created that provide the opportunity to correct errors related to:
  - Basic division facts.
  - Basic multiplication facts.
  - Omission of the decimal point.
  - Incorrect placement of the decimal point.

For example,

What do you think is wrong? Why do you think it is wrong?

1.6 ÷ 2 = 8 **0.15 ÷ 3 = 0.5** **5.6 ÷ 8 = 0.8**

- **Selecting Missing Parts of Equations**
  A divisor and a quotient are shown and the request is made to try to identify the correct placement of the decimal point in the dividend to make the equation true and to explain the strategy that was used to make the decision. The strategies are compared and calculations are shown to check the results.

Where does the decimal point belong to make the equations true? How do you know?

126 ÷ 3 = 4.2 **126 ÷ 3 = 0.042** **126 ÷ 3 = 0.42** **126 ÷ 3 = 42.0**

Assessment Suggestions

A multiple choice format can be used to gain insight into understanding if students are requested to defend the selection made and indicate, orally or in writing, why the other options were not selected or are inappropriate.

Possible Types of Assessment Tasks

Place the decimal point to show the answer.

8 x 0.34 = 272 | **0.142 x 3 = 426** | **14.2 x 3 = 426** | **1.42 x 3 = 426**

$0.25 ÷ 2 = 125 | **$2.73 ÷ 3 = 91** | **0.648 ÷ 4 = 162** | **48.5 ÷ 8 = 605**

Select the answer and explain why it is chosen.
Explain why you think the other choices are incorrect.

9 x 0.122 = □ 10.980 1.098 109.800
6.05 x 4 = □ 2.42 0.242 24.2
0.8 x 5 = □ 0.4 4.0 40.0

4.9 ÷ 7 = □ 7.0 0.7 0.007
5.75 ÷ 5 = □ 1.15 11.5 0.115
17.4 ÷ 3 = □ 0.58 5.8 58.0
0.294 ÷ 6 = □ 0.409 0.49 0.049

Where does the decimal point go to show true equations?
How do you know?

174 x 5 = 87.20 | **174 x 5 = 8.720** | **174 x 5 = 0.872**
1632 ÷ 8 = 2.04 | **1 632 ÷ 8 = 0.204** | **1 632 ÷ 8 = 20.4**

Look at each equation.
Will the answer be greater than one, less than one or about one?
How do you know?

0.096 ÷ 8 = □ | **0.96 ÷ 8 = □** | **9.6 ÷ 8 = □** | **96.0 ÷ 8 = □**

Explain your strategy for estimating the quotients:

6.40 ÷ 2 = □ | **0.64 ÷ 2 = □** | **0.064 ÷ 2 = □?**

What do you think is wrong? Why is it wrong?

5 x 5.04 = 2.552 | **2.01 x 4 = 0.801** | **3.5 ÷ 7 = 5** | **4.05 ÷ 5 = 8.1**

Reporting

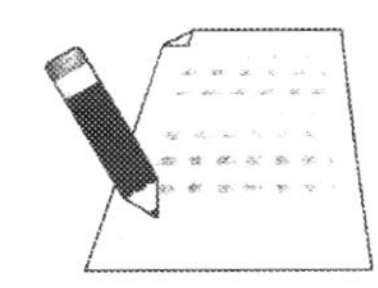

The responses collected from the assessment items make it possible to share information about the understanding of multiplication and division involving decimals that can include statements about ability to:

- Estimate products and quotients.
- Explain whether or not estimates are greater or less than a product or quotient.
- Look at equations and tell whether or not quotients are greater or less than one.
- Tell where a decimal point should appear in an answer and explain why it belongs in that position.
- Identify and correct errors related to computation procedures and in the placements of the decimal point.

For Reflection

1) What examples from teaching about the multiplication and division with decimals could you use to illustrate that *number sense* is the key foundation for *numeracy*?

2) How would you respond to someone who thinks that too much time is spent on activities that involve the estimation of answers and the comparison of estimation strategies?

3) How would you respond to someone who makes the claim that mathematics is a set of rules and it is very efficient to explain the rules for carrying out computations and then move on to new procedures and skills?

4) What are some possible advantages and disadvantages of assessing students by requesting the answers for twenty equations of the same type, i.e., multiplication with decimals?

5) What are some possible advantages and disadvantages of having students complete as many equations of the same type, i.e., division with decimals, in a timed test setting?

Addition and Subtraction with Fractions

Rules, Rote Learning – About Fractions

A research report about fractions published some time ago[1] included the following summary statement,

Almost all children appeared to search their memories for rules
and then attempted to apply them.
Although these rules were often misapplied, the students could not tell
that this had happened.
Of those who did use rules correctly, few had an internal conviction
that the rules were accurate or sensible (p.64).

The responses collected during diagnostic interviews with students and adults over several decades indicate that this confusion still exists. It is fascinating to see that almost all subjects know that for the division of fractions the rule is, *invert and multiply*. It is very rare, indeed, to come across someone who knows why that works. Some subjects invert the dividend rather than the divisor.

The results from one research project[2] indicate that under half of the American elementary teachers included in the study performed the calculations for $\mathbf{1\frac{3}{4} \div \frac{1}{2} = \square}$ correctly and only one was able to come up with a technically acceptable, but pedagogically questionable, story problem. Data collected from teachers-to-be over many years support the results from this report. Many students and adults are unable to make up a meaningful story problem for an equation like $\mathbf{8 \div \frac{1}{2} = \square}$ and they will suggest that calculating the answer for this equation is equivalent to solving *'eight divided by two'* ($\mathbf{8 \div 2 = \square}$) or *'one-half of eight'* ($\mathbf{\frac{1}{2} \times 8 = \square}$).
The question that needs to be asked is,

What is a possible reason for this lack of understanding and confusion?

A common error pattern for addition has subjects add both the numerators and the denominators. This was the case for one student from the sixth grade who recorded $\frac{2}{6}$ as the answer for $\frac{1}{4} + \frac{1}{2} = \square$. The interview strategy of illustration was employed to determine whether or not the student was able to correct his thinking and derive a rule for adding fractions. Blocks of the same size were made available and the student responded to the request, *show me how you could calculate the answer with these blocks and talk to me as you are working out the answer.*

I have four pieces, so that's like four-fourths. I take one piece, that's one fourth,
then I put it together with two pieces,
that's one-half and it makes three pieces, that's three fourths.

The next obvious step was to attempt to see whether or not the student could generalize and come up with a rule for what he has described. Such a rule may make it possible to correct the earlier mistake. However, as happens during interviews, answers to questions can be unpredictable and a little surprising.

So what should you always do when you add one-fourth and one-half?
'Always use blocks.'
A unique and unexpected generalization!

A girl from grade eight added and subtracted the numerators and denominators. Her explanations for calculating the answer were consistent for several examples and they included more than a little surprise.

According to this girl,

$\frac{9}{10} - \frac{5}{10} = 4$ *because* $9 - 5 = 4$ *and* $10 - 10 = 0$ *and* $\frac{4}{0} = 4$.

$3\frac{7}{10} - \frac{3}{10} = 34$ *because* $3 - 0 = 3$ *and* $7 - 3 = 4$ and $10 - 10 = 0$ *and* $\frac{4}{0} = 4$.

The manner of her delivery of the explanations left no doubt that she was convinced that she was using the correct procedure.

Without the presence of fraction *number sense* and without *conceptual understanding* operations with fractions are learned in a rote procedural way. Rules that are to be followed are presented and this is followed by extensive repetitive rote practice. According to the definition adopted in this book, some of the major shortcomings or weaknesses of this method of 'teaching' are illustrated by the results from the studies that were quoted and by the interview responses collected from different subjects who seemed to have experienced this type of teaching. Books on error patterns include many pages of examples from operations with fractions.

An advertisement for a workshop includes the description of introducing the addition of fractions to students in grade two.[3] A rule based delivery of skills followed by extensive practice is advocated. Two rules are introduced to students for the addition of fractions. One rule tells students what to do when like fractions, or fractions that have the same denominator are added. The other rule tells what to do to make the denominators the same. The description includes the claim that the procedure of introducing rules and providing extensive practice can result in having students in grade two *learn to perform operations with fractions flawlessly in less than a month* (p.13). The framework and content of this book can be used to generate the following questions about this approach or any similar rule based approach:

- *Why spend one tenth of a grade two student's mathematics program learning how to perform operations with fraction by rote? Is it possible that there are more valuable skills, procedures and ideas that could and should be learned by students in this grade?*

- *At this grade level, what does the knowledge of addition with fractions connect to in terms of previous learning; ongoing learning; what is learned next; or related experiences outside the classroom? Are these students able to visualize the aspects of the calculations they carry out?*

- *How is 'flawless learning' defined and assessed? What is done when, according to the definition, the learning is not flawless? Is it assumed or is there some sort of evidence that this flawlessness is permanent? What happens should a rule be forgotten or be confused with a rule for a different type of task?*

This brings to mind a television special about the scientist Sir Isaac Newton who also taught mathematics. The actor who portrayed the scientist uttered the statement, *you can teach any blockhead to repeat, but to make him understand is a challenge.* Over the years many teachers-to-be have bemoaned the fact that they went through the system without acquiring any sort of understanding. Frequently credit was given to teachers who managed to coach them to pass tests. It is possible to make students test-wise, but that does not mean that they are numerate.[4]

Some of the examples with fractions that are described go beyond addition and subtraction. The reason for the inclusion is two-fold. Possible disadvantages of an emphasis on rote rule learning are identified and at the end of this section these examples are used to illustrate the power of *number sense*, *conceptual understanding* and ability to *visualize*.

Pre-requisites – Key Ideas from the Previous Grades

The pre-requisites for the ability to develop *personal strategies* for the addition and subtraction of fractions include:

- *Conceptual understanding* of addition and subtraction.
- Knowledge of the basic addition and subtraction facts.
- Knowledge of the basic multiplication and division facts.
- Fraction *number sense* which includes the ability to:
  - *Visualize* fractions. When **a** over **b** is a fraction, students can think of and 'see' **a** out of **b** pieces or equal parts, where equal means congruent.
  - *Think flexibly* about fractions. Students know that any given fraction can be named in many different ways and they can generate their own rules for writing different names for a fraction.
  - *Estimate* fractions. Students can classify fractions less than one whole into categories: close to zero, close to one-half and close to one whole. Students can use the benchmarks **0**, $\frac{1}{2}$ and **1** as they estimate.
  - *Relate* fractions. Students can compare any two fractions and put three or more fractions in order from least to greatest or vice versa – like fractions (same denominator), unit fractions (one as a numerator), and unlike fractions (different numerators and denominators)
  - *Connect* fractions. Students know who uses fractions, when, were and why.

Addition with Fractions

Observations during introductory lessons to grade four and grade five students who had spent time on the pre-requisites ideas and skills support the conclusion that rules do not need to be presented during an introduction. After meaningful problems were described, equations were recorded, answers were calculated and recorded and rules for the addition of like as well as unlike fractions were generated by the students. Such a sequence of activities is an example of learning new ideas *through problem solving.*

Addition of Like Fractions

Several problems of the following type can be presented:

If two people each give you one-fourth of an orange,
how many fourths will you have?
If each of three people gives you one-third of a chocolate bar,
how many thirds will you have?

The request is made to record equations and the answer for each problem.
After the equations and the answers are compared, the students are invited to,

Make up a rule for calculating the answers when fractions like these or fractions that have the same denominator are added.

A non-example could be part of the introductory setting.
For example, the students are told,
Someone solved the equation $\frac{1}{2} + \frac{1}{2} = \square$ *by recording* $\frac{2}{4}$ *as the answer.*
How do you think that person calculated that answer?
How do you know that the answer is incorrect? Try to think of two different reasons.
Why does it not make sense to add the denominators?

Some guidance may be required during the follow up discussion to have students realize that the answer is equal to one of the addends and that when the additive action is performed, the size of the pieces that are added, as indicated by the denominator, does not change.

Addition of Unlike Fractions

The request is made to record an equation and to try and use a sketch that would help in recording the answers for the equations for each of the following problems.
After eating one-half of one pizza, the family ate one-fourth of another pizza.
How much pizza did the family eat?

For dessert there were two different pies of the same size.
The family ate one-third of one pie and one-half of the other pie.
How much of the two pies did the family eat?

The equations for the problems are recorded and displayed:
$\frac{1}{2} + \frac{1}{4} = \square$ $\frac{1}{2} + \frac{1}{3} = \square$
What were you able to tell about the answers by just looking at your sketches?

With some guidance, the students will be able to share the observations that the answers are greater than one-half and less than one whole. If for some students their sketches led them to conclude that the answer for the first equation is three-fourths, the following questions are posed,
How might it be possible to calculate the answer three-fourths?
Why is it not easy to tell the answer for the second equation?
What could be done to make it easy to calculate the answer for the second equation?

The students are led to the conclusion that other names for fractions are needed to make the denominators the same. When this is done, the rule for adding fractions that has been generated can be used. As part of the procedure to write other names for fractions, the generalization for *flexible thinking* about fractions that was an important aspect of developing fraction *number sense* may have to be revisited.

Subtraction with Fractions

Since the generalizations for subtraction and addition with fractions are very similar, the strategies and procedures that were suggested for the addition can be used.

Subtraction of Like Fractions

Several equations can be shown on the chalkboard.

$\frac{7}{10} - \frac{5}{10} = \square$ $\frac{3}{4} - \frac{1}{4} = \square$ $\frac{2}{3} - \frac{2}{3} = \square$

The requests that are made can include,

Make up a story problem for each equation.

It is unlikely, but if hints are required these could ask students to think about food.

Record the answer for the problem.

Make up a rule for calculating the answer for subtracting fractions.

The request is made to record an equation for the problem, to solve the problem and to be ready to explain the strategy that was used.

There are two pizzas in the fridge. You were allowed to eat one-third of one pizza.
How much pizza will be left in the fridge?

During diagnostic interviews it is easy to identify responses from students that are indicators of lack of fraction number sense. Rules for performing operations are recited. Subjects are unable to make a prediction about an answer or make a comment about the reasonableness of an answer. The explanation of a solution procedure for $\mathbf{2} - \frac{1}{3} = \square$ suggested by a student in grade five during an interview illustrates this dilemma.

First you write the two as two over one.
Then you change that to thirds and write six over three.
Then you subtract one-third.
That leaves five-thirds or one and two-thirds.

Results from interviews and conversations with adults indicate that the majority would 'cut up both pizzas into thirds' to calculate the answer. There is no connection to anything that was done previously with subtraction. Common rationales for this action include, '*That is how it is done*' and, '*That is how I was taught.*'

Subtraction of Unlike Fractions

It is easy to present a problem that leads students to the realization that the subtractive action to determine an answer to a problem cannot be simulated when the denominators of fractions are different. Once the denominators are the same, this action can be simulated with objects or shown with a sketch which makes it possible to record answers for equations. These recordings can then be used to create a personal rule for calculating answers when fractions are subtracted.

The request is made to make up story problems for several equations.
For example,

$\frac{3}{4} - \frac{1}{2} = \square$ $\frac{7}{10} - \frac{3}{5} = \square$ $3 - \frac{3}{4} = \square$ $\frac{5}{6} - \frac{1}{3} = \square$

Questions of the following type can be part of a discussion as the equations are examined,

Are there any equations that you know the answer for just by looking at the equations?

If that is the case, *explain your thinking.*

Why is it not easy to recognize the answers for each equation?

What do you think would make it possible and easy to show and explain how to calculate the answers for each of the equations?

- **Accommodating Responses – Use of Language**

During diagnostic interviews that deal with fractions *number sense* and operations with fractions, many subjects use the expression *'reduce'* or *'reduce the fraction.'* Some subjects will share the fact that their answers are marked wrong if *we don't reduce them.*

The use of these expressions is detrimental to fostering *visualization* since it implies the opposite of what actually is taking place. What happens when $\frac{4}{6}$ is changed to $\frac{2}{3}$? If fractions are *visualized* in terms of congruent parts of regions, there are fewer pieces or the number of pieces is reduced, but the size of the pieces has increased. Since that is the case, it is advantageous for students to use expressions that can enhance the *visualization* of this result. If students are requested to change fractions to simplest form, they should be able to state in their own words what that means. Such a statement should describe the characteristics of fractions in simplest form.

The Power of Number Sense and Conceptual Understanding

The results from many diagnostic interviews and conversations with adults support the conclusion that learning about operations with fractions in a via rule and rote practice setting is detrimental. Rules are not only forgotten, but without understanding they are also applied incorrectly. *Number sense* along with *conceptual understanding* is the key foundation for *numeracy*. Examples of multiplication and division with fractions are used to illustrate this conclusion.

Students' knowledge of the *groups of* interpretation of multiplication and their fraction *number sense* enables them to derive their own generalizations or rules. They are able to record equations for problems, record the answers for the equations and from the answers come up with a rule for multiplying whole numbers with fractions and fractions with fractions.

A request is made to record equations for word problems that are presented.

During one break, each of the five basketball players ate one-half of an orange.
How many oranges were eaten? $5 \times \frac{1}{2} = \square$

The students who record $\frac{1}{2} + \frac{1}{2} + \frac{1}{2} + \frac{1}{2} + \frac{1}{2} = \square$ require a brief reminder about another way of writing addition equations of this type.

How much is one-half of a group of one-half of a pizza? $\frac{1}{2} \times \frac{1}{2} = \square$

For the students who record $\frac{1}{2} \div 2 = \square$ the question of calculating the area of a $\frac{1}{2}$ by $\frac{1}{2}$ rectangular region could be illustrated with a sketch and the *partitioning* or *equal sharing* interpretation of division could be simulated to explain the difference between the two equations.

The recording of the answers $\frac{5}{2}$ and $\frac{1}{4}$ leads to the request for a rule for multiplying whole numbers with fractions and fractions with fractions. This rule can then be tested with further examples of problems and equations.

What are possible reasons for the outcome that almost one-half of twenty three American elementary teachers in a study[(2)] incorrectly performed the calculations for $\mathbf{1\frac{3}{4} \div \frac{1}{2}} = \square$ and only one was able to come up with an acceptable story problem? Somewhere along the way, these teachers were 'taught' the rule for dividing fractions. This result illustrates that rote learning does not make it possible to retrieve something that is forgotten nor to know whether or not a rule has been applied correctly. Lack of *conceptual understanding* of the operation does not make it possible to *visualize* the answer without performing any pencil and paper calculations.

Visualization is enhanced by knowing the two interpretations of division. When the divisor is a name for a whole number, i.e., $\frac{3}{4} \div \mathbf{3} = \square$, the *equal sharing* interpretation of division makes it possible to estimate and make a prediction about the quotients. When the divisor is a fraction, as is the case for the example from the study, the *equal grouping* interpretation makes it possible to estimate and predict the quotient.

The ability to *connect* can be thought of as a problem solving strategy. The *equal grouping* interpretation of division enhances the ability to *connect*. For example, the equation from the study, $\mathbf{1\frac{3}{4}} \div \frac{1}{2} = \square$, could be interpreted as,

How many bags holding $\frac{1}{2}$ pound can be filled from $\mathbf{1\frac{3}{4}}$ pounds?

The quotient can be *visualized* by thinking: Three groups of one-half can be taken or three bags can be filled with $\frac{1}{2}$ pound each and $\frac{1}{4}$ will be left to fill another bag half full. After $\mathbf{3\frac{1}{2}}$ is recorded as the quotient, the next challenge would be to try to come up with a calculation procedure that yields this answer.

When results from action research or observations cite the conclusion that a group or groups of students lack *conceptual understanding* or lack *number sense*, the references that were part of the students' mathematics learning need to be examined. Without a focus or an emphasis on these important aspects of mathematics learning such results should not be surprising to anyone. At present, or at the time of writing, many student references lack this emphasis. Appropriate references are essential along with professional development that shares key strategies required for reaching the goals advocated in these references. The assessment strategies suggested in references need to be re-examined and may require a new focus. Determining whether or not mathematics is meaningful to students requires a focus on basic ideas and not just basic skills.

Appropriate Practice

Types of Activities and Problems

- **Missing Fractions and Missing Parts of Fractions**

What is missing? How do you know? Explain your thinking.
Try to use a diagram or objects to illustrate your answer.

$\frac{2}{3} + \frac{\square}{\square} = 6$ $\quad$ $\frac{2}{3} + \frac{\square}{3} = 4$ $\quad$ $\frac{\square}{4} + \frac{1}{2} = \frac{3}{4}$ $\quad$ $\frac{\square}{10} + \frac{\square}{\square} = 1$

$\frac{5}{6} - \frac{\square}{6} = \frac{1}{2}$ $\quad$ $\frac{7}{10} - \frac{\square}{\square} = \frac{1}{5}$ $\quad$ $4 - \frac{\square}{2} = 2\frac{1}{2}$ $\quad$ $\frac{\square}{5} - \frac{\square}{5} = \frac{\square}{5}$

- **Finding and Correct the Mistakes**

  What is wrong? Why do you think the mistakes are made?
  How would you correct the mistakes?

  $\frac{1}{2} + \frac{1}{2} = \frac{2}{4}$ $\quad \frac{2}{3} + \frac{1}{6} = \frac{3}{9}$ $\quad \frac{1}{4} + \frac{1}{3} = 12$ $\quad \frac{8}{10} + 1 = \frac{8}{10}$

  $\frac{2}{3} - \frac{1}{3} = \frac{1}{6}$ $\quad 2 - \frac{1}{2} = \frac{1}{2}$ $\quad \frac{3}{10} - \frac{3}{10} = 1$ $\quad 4 - \frac{1}{4} = 1$

- **Following Directions**

  Several digits and instructions to be followed are presented.
  For example,

  Use the digits ***1***, ***2***, ***3***, *and* ***4***.

  For the equation $\frac{\square}{\square} + \frac{\square}{\square} = \frac{\square}{\square}$ try to determine:
  - *The greatest possible sum.*
  - *The least possible sum.*
  - *A sum between the greatest and least possible sum.*

  During the follow-up discussion responses are compared and strategies are explained.

  For the equation $\frac{\square}{\square} - \frac{\square}{\square} = \frac{\square}{\square}$ try to determine:
  - *The greatest possible difference.*
  - *The least possible difference.*
  - *A difference between the greatest and least difference.*

  During the follow-up discussion responses are compared and strategies are explained.

- **Different Equations**

  Try to think of at least two different equations for each of the answers.

  $\frac{\square}{\square} + \frac{\square}{\square} = \frac{3}{4}$ $\quad \frac{\square}{\square} + \frac{\square}{\square} = \frac{9}{10}$ $\quad \frac{\square}{\square} + \frac{\square}{\square} = 1$

  $\frac{\square}{\square} - \frac{\square}{\square} = \frac{1}{2}$ $\quad \frac{\square}{\square} - \frac{\square}{\square} = 1$ $\quad \frac{\square}{\square} - \frac{\square}{\square} = \frac{1}{4}$

  During the follow-up discussion responses are compared and strategies are explained.

Assessment Suggestions

Possible Types of Assessment Tasks

- Calculate the answers:

  $\frac{3}{10} + \frac{7}{10} = \square$ $\frac{1}{3} + \frac{1}{6} = \square$

  $\frac{3}{4} - \frac{1}{4} = \square$ $3 - \frac{1}{3} = \square$ $\frac{2}{3} - \frac{1}{4} = \square$

  Create a problem for the equations.
  Provide oral or written responses for:

  How do you know your answers are correct?
  How can you show in some way that your answers are correct?

- The request is made to try to correct the mistakes.

  $\frac{1}{4} + \frac{1}{2} = \frac{2}{6}$ $\frac{3}{10} + \frac{1}{5} = \frac{4}{10}$ $\frac{3}{4} - \frac{1}{3} = \frac{2}{1}$ $2 - \frac{1}{2} = \frac{1}{2}$

  How do you know the answers are incorrect?
  How could you tell the answers are incorrect just by looking at the equations?
  What are the mistakes? Why do you think the mistakes were made?
  What are the correct answers?

- The following requests are made:

  What is your rule for adding fractions? Illustrate your rule with an example.
  What is your rule for subtracting fractions? Illustrate your rule with an example.

Reporting

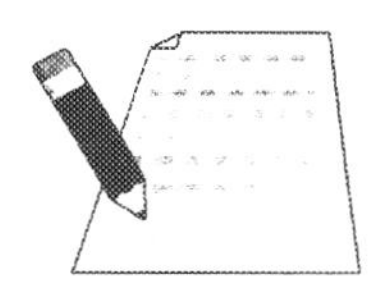

The observations and assessment items make it possible to share information about:

- Ability to *generalize*. Were the rules clear and correct?
- Ability to *connect*. Were the story problems that were created meaningful?
- *Confidence and willingness to take risks.*
- Ability to *visualize* and *estimate*: Was the interpretation of the actions for the operations correct and could predictions be made about answers.
- Indicators of *fraction number sense* which can include:
  - *a) Visualization* of fractions – Could appropriate sketches be drawn?
  - *b) Flexible thinking* about fractions – Could different names be used for a fraction?
  - *c) Estimating* strategies – Was reference made to the use of *benchmarks*?
  - *d) Relating* fractions – Were there indicators of ability to compare or order fractions or place these on a number line?
- Indicators of *confidence* and *willingness to take risks.*

For Reflection

1) Do you think students who do not change their answers to the simplest form should get these marked as incorrect?

2) How would you respond to someone who makes the point that since all fractions can be changed to decimals, students should not be bothered with having to learn how to perform operations with fractions?

3) How would you respond to someone who suggests that calculators that perform operations with fractions make any time spent on learning how to perform operations with fractions redundant?

4) Someone makes the point that since students in grade six are capable of thinking like adults they do not need manipulative materials to learn about the operations with fraction. According to this person, rules, explanations and examples presented to students should suffice. How would you respond to this person?

5) The title of an article in the newspaper[(5)] stated, **Public School Teachers Drop Grades From Evaluations** and continued with, *focus has shifted to what they were learning.* The report included the following information about teachers in one city's school system, *Under a sweeping policy review underway, many teachers have stopped giving students grades of less than* ***50%*** *- or have removed grades entirely from evaluations of students' work. It's all part of a trend toward a more "creative" approach to judging school work.*

   According to the report, one math teacher who taught underachieving students in high school got rid of grades three years ago because, *A lot of kids were getting the message they weren't any good because they got a* ***65*** *or whatever.*

   Whether you agree or disagree,
   what questions would you like to ask teachers who get rid of marks?
   What are some possible advantages and disadvantages of getting rid of grades?
   What is your conclusion about giving grades?

   How do you think the teacher who was quoted might define and determine a measure of *underachieving*? How would you define this term? What examples would you consider to be possible indicators of *underachieving*?

6) What would you say to someone who claims that *more practice will eventually lead to an understanding of or a better understanding of what is practiced?*

Problem Solving with Percent

Percent *number sense*, which includes the ability to represent percents on a **10** by **10** grid model similar to the one hundred base ten block or on an unmarked number line, makes it possible for students to *estimate* answers or to check the reasonableness of calculations with percent. *Connecting* to previous learning, which includes ideas about rate ratios, decimals, and the multiplication of decimals enables students to calculate percentages of numbers.

Pre-requisites – Key Ideas from the Previous Grades

The key ideas, procedures and skills about percent *number sense*, decimals, fractions and multiplication with decimals include:

- Ability to represent given percentages by shading parts of rectangular regions. For example: **50%, 25%, 75%**, about **10%**, about **90%**.
- Ability to use benchmarks to record estimates for shaded parts of rectangular regions.
- Ability to record equivalent fractions and ratios.
- Ability to connect fractions to decimals and to percent or vice versa: For example,
  25% means **25 out of 100** or twenty five hundredths – $\frac{25}{100}$ or **0.25**
  0.75 is equal to $\frac{75}{100}$ or **75 out of 100** or **75%**
  $\frac{1}{2}$ is equal to $\frac{50}{100}$ or **50 out of 100** or **50%**
- Knowledge of a personal strategy for multiplication with decimals.

Estimating and Finding Percent of a Number

Calculating the percent of a number is a common type of problem that is encountered in everyday situations related to paying taxes of different types or issues related to shopping such as discounts.

The list of pre-requisites indicates the power and importance of *connecting* previous learning to ongoing learning. The goal is to *connect* to previous learning and to create activities and problems that will enable students to come up with a personal strategy for calculating the percent of a number.

An introductory setting could include discussions about general estimation strategies.
For example,

*What is a strategy for estimating **10%** and **5%** of an amount of money?*

For example, **10%** and **5%** of **$2.50**
10% and **5%** of **$12.00**

*How would you estimate **10%** of an amount of money?*
*How can an estimate for **10%** be used to arrive at an estimate for **5%**?*

What percentages do you think are easy to estimate? Why is that the case?
Use examples to explain your thinking

A problem is presented,

*An advertisement for a store offers a **25% reduction** in prices.*
*Someone plans to buy something that cost **$60.00** before the reduction.*
How much money will be saved?

The estimation of the answer can begin with an unmarked number line that shows only **0** and **60**. The discussion results in:

Placement of a mark on one-half or **50%** of **60** or **30** of the number line.
Then one-half of **50%** or **25%** or **15** is identified.

The calculations that are carried out have to result in an answer or a saving of **$15**.

25% of **$60** is the same as $\frac{25}{100}$ of **$60** or **0.25** of **$60**.

The *of* is an abbreviation for *groups of* which means that **0.25** *of* **$60** can be rewritten

as **0.25** x **$60** = □.

$$\begin{array}{r} \$60 \\ \times\ 0.25 \\ \hline 300 \\ 1200 \\ \hline 1500 \end{array} \text{ or } \$15.00$$

The list of easy to estimate percentages that was generated should include **50%, 25%** and **75%**.

*How could estimates for **50%, 25%** or **75%** be used to arrive*
*at estimates for **13%**, **60%** or **85%**?*
*Try to think of at least two ways to estimate **90%** of an amount of money.*

Estimating and Finding Percent

The problem of determining what percent one number is of another number involves the recording of a matching fraction or ratio with denominator one hundred or as percent.
For example,

*If **14** out of **20** basketball free throws are successful,*
what percentage of the throws is successful?

An unmarked number line showing **0** and **20** can be used to illustrate an estimation strategy. The discussion results in:

- Recording **100%** below the **20**.
- Identifying the half-way mark or **10** as **50%**.
- Identifying the half-way mark between **10** and **20** or **15** as **75%**.
- The conclusion: Since **14** is close to **15**, the estimate is about **75%**.

The calculation or the determining of the answer involves:

- Recording **14** out of **20** as the ratio $\frac{14}{20}$.
- Changing the ratio to out of one hundred or, $\frac{70}{100}$.
- Recording the ratio $\frac{70}{100}$ as a decimal **0.70** and then recording it as **70%**.

After several problems of this type have been solved, students could be reminded that the ratio **a/b** can also be interpreted as **a ÷ b**. When the division interpretation is entered into a calculator, a decimal is displayed. The multiplication of the decimal by **100** changes the ratio to percent.

Appropriate Practice

Types of Activities and Problems

Record the Missing Numerals

Use sketches or a number line to explain your thinking.

If **100%** is equal to **$60**, determine the following amounts of money:

50% = □ **25% = □** **75% = □** **10% = □** **60% = □** **150% = □**

If **25%** is equal to **$20**, determine the following amounts of money:

100% = □ **75% = □** **50% = □** **About 10% = □** **125% = □**

Record the Missing Percentages

Use sketches or a number line to explain your thinking.

If **$30** is equal to **25%,** determine the following percentages.

$60 = □% **$12 = □%** **$90 = □%**

$120 = □% **$48 = □%** **$240 = □%**

Estimating Percentages

Use sketches and explain your strategies.

$\frac{7}{8}$ **is about □%** $\frac{4}{9}$ **is about □%**

$\frac{22}{66}$ **is about □%** $\frac{21}{19}$ **is about □%**

Connecting

Where can percent signs (**%**) be seen? Make a list. Use sketches to illustrate the meanings of the percentages on your list. Compare the lists and sketches.

Assessment Suggestions

Possible Types of Assessment Tasks

a) Use a sketch and explain your strategy, orally or in writing, for determining:

25% of **44** **90%** of **60**

b) If **20%** is equal to **$15**, explain orally or in writing, how you would determine the amount of money for **100%**.

c) Use sketches to show how you would estimate:

The percentage for **13** out of **20** free throws.

The amount for **30%** off for something that costs **$80.00**.

d) Four out of five kittens in a litter are black.

What percentage of the kittens is black? Explain your thinking.

e) **45%** of the students in a classroom are girls.

Record your answers for,

The percentage of the students that are boys.

How many girls and how many boys could there be? Explain your thinking.

f) Where and how are percents used? Write a sentence.

Reporting

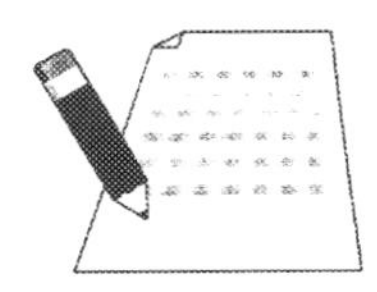

The types of assessment tasks can provide indicators of ability to:

- Calculate percentages: As calculation strategies are explained and illustrated, indicator of ability to *visualize* percentages may be noted.
- Estimate percentages: Estimation strategies will yield information about the *benchmarks* that are used. The ability to use more than one estimation strategy is an indicator of *flexible thinking*.
- Explain Strategies: The sketches and explanations will provide information about ability to *communicate mathematically*.
- *Connecting*: Are students aware of why percentages are discussed?

Addition and Subtraction with Integers

The intent is to show, as was the case for the previous topics, that integer *number sense* and *conceptual understanding* of the operations will enable students to develop personal strategies along with rules for performing the operations.

Pre-requisites – Key Ideas from the Previous Grades

A key pre-requisite is related to *visualization*. Different models, representations and ideas can be used to foster the ability to visualize. These can include:

- Objects of different colours to indicate positive and negative charges.
- Signs to indicate money earned and money spent, and defining net worth.
- The number line or the number line that is part of a thermometer.
- Plus and minus statistics that are part of sports like hockey, football and golf.
- Making references to above or below sea level.

Relating integers. This includes the ability to compare and order integers.

Flexible thinking about integers. The definition of net worth can be used to represent a given integer in different ways. For example, a net worth of **$2** or a positive two could mean having three and owing one or having four and owing two. Earning one more dollar while spending one more dollar does not change the net worth. The same idea can be illustrated with a positive and negative charge model. Adding another positive and negative charge is equivalent to adding zero, **1 + -1 = 0.**

Addition with Integers

The presence of integer *number sense* and *conceptual understanding* of addition will enable students to calculate sums for integer addition equations and to create a rule for performing the operation.

Problems of the following type are presented:

*Someone earned $**4** and spent $**6**.*
*Someone earned $**8** and spent $**10**.*

The initial question that should be addressed is,

What would make it possible to spend more than what is earned on any given day or at any one time?

The request is made to translate the problems into addition equations that can be solved to show the net worth for that day.
After **4 + -6 = □** and **8 + -10 = □**, respectively, are agreed upon and the answers are recorded, a comparison could be made to addition equations and answers learned previously.

Look at the first addend or numeral and compare it to the answer.
How can this comparison be described?
How does this result differ from previous addition equations?

Some guidance may be required before the students will realize that prior to this, answers were always either greater than the first addend or the same (**a + 0 = □**), but now the answers can be less than, equal to or the same as the first addend. After this conclusion about the three possibilities is reached, the students are invited to make up a problem and record equations for each of the possible scenarios:

- The answer is less than the first numeral.
- The answer is the same as the first numeral.
- The answer is greater than the first numeral.

The students could be challenged to use examples other than money.

The request is made to solve several equations of the following types and to make up a matching problem for each:

a) **-2 + 6 = □**
5 + - 4 = □
- 9 + 10 = □

b) **-5 + - 8 = □**
-3 + - 3 = □
-2 + - 6 = □

After the answers are recorded, the similarities and differences of the equations in each column are discussed.

The request is made to,

Make up a rule for calculating the answers when the signs are different.
Make up a rule for calculation the answer when the signs are the same.

The rules that are created are compared.

Appropriate Practice

Types of Activities and Problems

- **Record and Solve Equations for Problems**
  For example,
  You owed two dollars and borrowed ten more.
  You earned six dollars and spent eight dollars.

- **Create and Solve Problems for Equations**
  Try to think of and describe two different settings for each equation.
  For example,
  -5 + - 5 = ⎕ -6 + 4 = ⎕ - 7 + - 7 = ⎕ 4 + - 10 = ⎕

  During a follow up discussion, the different settings that were created are compared. What other settings might be possible?

- **Find the Mistakes**
  What do you think is incorrect?
  -2 + 2 = 4 6 + 6 = 0 7 + -2 = - 9 -4 + -4 = - 4 0 + 5 = -5

  Try to think of at least two different ways that will result in each of the answers that are shown.

- **Record the Missing Integers**
  Identify and record the missing integers.
  -5 + ⎕ = 5 ⎕ + -3 = -3 -8 + ⎕ = 0 ⎕ + ⎕ = -1
  Explain how it could be shown that your choices are correct?

Subtraction with Integers

At the conclusion of one session a young teacher in training wanted to know the rule for subtraction integers because the sponsor teacher had assigned that topic to him for the coming week. During a brief discussion it became evident that he was unable to retrieve or reinvent the rule that he at one time had known.

What is the point of this little true scenario? Memorized rules can be and are forgotten no matter how many times they are applied in practice. If that is the case, they are gone.

What can the young man who requested a rule do to lead his students to come up with the rule that he was looking for, *to subtract a negative integer, add its opposite*?

Conceptual understanding of subtraction and modelling can be used to create meaningful story problems and construct and record the answers for equations with a minuend less than or equal to the subtrahend. The ability to make up meaningful problems enhances the ability to *visualize* the answers for the equations.
For example,

-6 – - 4 = □ **-8 – -8 =** □ **-5 – - 2 =** □

The answers of **-2**, **0** and **-3** may enable some students to try to create a rule for calculating the answer when a negative integer is subtracted from a negative integer. Should that be the case, this rule can be tested for equations with a minuend greater than the subtrahend.
For example,

-2 – -5 = □ **-7 – -10 =** □ **-1 – -2 =** □ **-3 – -7 =** □

The key to modelling the equations is the ability to *think flexibly* about integers. This aspect of *number sense* enables students to represent or model an integer in different ways.
For example, for the first equation, **-2 – -5 =** □**,** the challenge is to represent a negative charge of two; a debt of $**2;** or being minus two after a hockey game in a way which makes it possible to subtract a negative **5**. The ability to represent or to think of an integer in different ways is based on the knowledge that one, minus one or zero can be added to a net worth or value without changing it.
A debt of $**2**, or **-2**, for example can mean:

Owing $**3** or **-3** and having $**1**.
Owing $**4** or **-4** and having $**2**.
Owing $**5** or **-5** and having $**3**.

The last representation makes it possible to subtract a debt of $**5** or **-5** and it shows that the answer is a positive **3** or $**3**. Students can be reminded that this flexible thinking about integers or finding another name for an integer is similar to the thinking used for whole numbers and fractions, i.e., **52 – 27 =** □; $2 - \frac{1}{3} =$ □.

The remaining equations can be solved in similar fashion and the students can be challenged to make up a rule for subtracting negative integers.

Further guidance to reach the conclusion can be provided by recording the last four equations in a row and then recording related addition equations in a second row:

a) **-2 – -5 = [ ]** b) **-7 – -10 = [ ]** c) **-1 – -2 = [ ]** d) **-3 – -7 = [ ]**

a) **-2 + 5 = [ ]** b) **-7 + 10 = [ ]** c) **-1 + 2 = [ ]** d) **-3 + 7 = [ ]**

After the equations are solved a comparison between related equations in rows is made.

How are the members of the related equations different?
What is the same about the related equations?
If you are asked by someone for a rule to calculate the answer when two negative integers are subtracted, what would your rule be?

The challenge is presented to pairs of students to record equations with integers and to make up meaningful problems for each of the following scenarios:

The answer is less than the minuend or the first numeral in the equation.
The answer is equal to the minuend.
The answer is greater than the minuend.

After the equations and problems are shared, a comparison is made to previous learning.

Which of the equations is different from subtraction equations for numerals other than integers?
How is it different?

Appropriate Practice

Types of Activities and Problems

- **Record and Solve Equations for Problems**
  For example,
  You owed five dollars and paid back three dollars.
  You owed three dollars, earned five dollars and paid your debt.
  Use a sketch to show that the answers are correct.

- **Create and Solve Problems for Equations**
  Try to think of and describe two different settings for each equation.
  Record the answers.

  -8 – 5 = □ 6 – - 9 = □ -2 – -3 = □ -4 – -4 = □

  During a follow-up discussion, the different settings that were created are compared. What other settings might be possible?

- **Find the Mistakes**
  What do you think is incorrect?
  10 – -10 = 1 -8 – 0 = -7 -5 – 1 = -6 -8 – -2 = - 6

  Try to think of at least two different ways that will result in each of the answers that are shown.

- **Record the Missing Integers**
  Identify and record the missing integers.
  -3 – □ = -3 -9 – -10 = - 19 □ – -5 = -5 -□ – □ = -4
  Explain how it could be shown that your choices are correct?

- **Record the Missing Signs and Integers**
  Identify and record the missing signs and integers.
  -2 ⌑ -2 = -4 -2⌑ -2 = 0 -2⌑ □ = -2 -2⌑ □ = 2 □⌑ □ = -1
  Explain how it could be shown that your choices are correct?

- **Recording Equations for Instructions**
  Create equations for each of the following and keep a record of the different attempts:
  - Add two negative integers for the answer **-5**.
  - Add a negative and a positive integer for the answer **-5**.
  - Subtract two positive integers for the answer **-2**.
  - Subtract two negative integers for the answer **-2**.
  - Three different equations with an answer of **-4**.

  Equations, problems and strategies are compared.

Assessment Suggestions

Possible Types of Assessment Tasks

a) Make up story problems for the equations: **-3 + - 6 = □** **-4 – - 5 = □**

b) Calculate the answers and use sketches to show that the answers are correct:
-6 + 2 = □ **-2 – -4 = □**

c) What is wrong? **-3 + -3 = 0** **4 – -4 = 8**

d) Record equations for the answer **-3**:
- **(i)** Two integers are added.
- **(ii)** Two integers are subtracted.

e) Use examples to explain what you know about calculating the answers when:
- **(i)** Two integers that have different signs are added.
- **(ii)** Two negative integers are added.
- **(iii)** Two negative integers are subtracted.
- **(iv)** A negative integer is subtracted from a positive integer.

Reporting

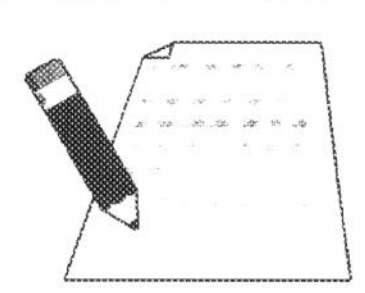

Responses to the assessment tasks can make it possible to share information about students' ability to;

- *Connect* addition and subtraction with integers to events from their experience by making up meaningful word problems for equations.
- Calculate sums and differences for the different possible combinations of integers and illustrate with sketches that show that the answers are correct.
- Identify and correct errors in calculation and use sketches as part of the explanations.
- *Think flexibly* by providing different answers for an equation with missing data and recording different equations for the same answer.
- Generalize by creating rules for different scenarios involving the addition and subtraction of integers and explaining the rules with examples.

For Reflection

1) How would you respond to someone who suggests that integers and operations with integers should not be part of the elementary mathematics curriculum?

2) How would you react to the suggestion that teaching the rules and then having students practice is still the most efficient way of teaching mathematics?

3) How would you react to someone who suggests that the estimation strategies for calculating percentages are not important because, '*I trust a calculator.*'

Chapter 6 – Measurement and Measurement Sense

About Measurement

There are a few key ideas that need to be the focus of activities that involve the measurement of different characteristics. Students will need to realize that there exists a similarity between the act of counting sets of discrete objects and the act of using measuring to describe parts of continuous quantity. Counting gives answers to *how many?* Measuring gives answers for *how much?*

One learning outcome is related to the use of correct language when comparisons are made or when results of measurements are discussed. The word *big* is inappropriate for most comparisons. It does not make sense to use the word *amount* with a description of discrete objects. Common incorrect usages by adults include making reference to *amount of people* and to *less people*.

The measurement of different characteristics will lead to the realization that the process of measuring is the same for each continuous quantity or for events. It is also true that the measurement skills and procedures that are learned and have been learned are the same for any system. However, the examination of relationships among different units that are used to measure the same characteristic is not the same for different systems.

About Measurement Sense

Measurement sense is not a common expression and it is not used in the curriculum. Since *measurement sense* involves visual imagery and visual thinking it is a component of *spatial sense*. Shaw and Puckett [(1)] identify four interrelated components of *measurement sense*. The first states that measurement sense implies knowing the units appropriate for a task and having formed useful mental pictures appropriate for a task. It also involves knowledge of the measurement process; having the ability to decide when to measure and when to estimate; and knowledge of several estimation strategies.

The classroom settings that are conducive to fostering the development of *measurement sense* require, as is the case for developing *number sense*, an emphasis on *conceptual understanding*. Settings are required that provide opportunity for students to *communicate mathematically*, orally as well as in writing about the key aspects of measurement. Many opportunities exist to compare estimation strategies and to explain the meaning and proper use of the instruments that are used.

Key Ideas from the Previous Grades

In the previous grades introductions to measurement included:

- Lengths and Distances – metre and centimetre.
- Capacity/volume – litre.
- Mass /weight – gram and kilogram.
- Time – second, minute, hour – digital and analog clocks.

The main skills, procedures and ideas from the previous grades can be summarized by examining a possible teaching sequence for measurement of length: [(2)]

- Definition of the Characteristic: The introduction and use of the appropriate terminology which includes: *as long as* or *same length*, *longer than* and *shorter than* to describe the characteristics of length as objects are sorted.

- Making Comparisons and Ordering: The use of appropriate terminology while comparisons are made (*long, longer, short, shorter*) and three or more objects are put in order (from *long* to *longest* or from *short* to *shortest*).

- Finding the answer to *How long?*
    - Selection of an arbitrary unit (body unit; paperclip).
    - Correct placement of the unit while measuring.
    - Measuring to the nearest unit and using *about* to report the result.
    - The use of a *referent* as part of estimation strategies.
    - Designing a measurement instrument – a ruler.
    - Explaining the meaning of the markings and numerals on a ruler.
    - Recognizing and explaining the need for a standard unit (centimetre).
    - Solving meaningful problems using the standard unit.
    - Estimating using the standard unit – referring to the referents used as part of the estimation strategies.
    - Explaining the need for a different standard unit (metre).

- Describing and using the relationship among the different units.

- Key ideas about the convention of using metric units when recordings are made:
    - Use of the 're' spelling.
    - Using abbreviations only with numerals.
    - Using the abbreviations without periods and without adding 's' for the plural form.

General Goals for the Intermediate Grades

The general outcome for measurement in the curriculum states that *students use direct and indirect measurement to solve problems*. Pre-requisites for the ability to solve problems include skills, procedures, ideas and generalizations related to measurement of length, area, capacity and volume.

The teaching sequence for length from the previous grades illustrates important skills and ideas that are learned and need to be learned for other characteristics. Some important ideas from this sequence may need to be revisited and then included as new topics are introduced.

Many important ideas related to aspects of measurement are well suited for developing new ideas *through problem solving*. Activities and problems dealing with aspects of measurement topics lend themselves to use imagination. Students could be asked to pretend to have lived a long time ago. How did these people solve problems that required making and reporting results of measurements of different quantities? Skits or pantomimes can be created to illustrate such journeys back into time.

Measurement of Length and Distance

Types of Activities and Problems for Key Ideas

Requests are made to prepare oral or written reports for:

- Who measures lengths and distances? When? Where? Why?
  When might measurement of length or distance be discussed in your home?

- Who is interested in measuring lengths or distances to the nearest centimetre or to the nearest metre?

- Describe two different strategies for estimating one side of a picture frame in centimetres and the length of the classroom in metres.

- How would you explain to a young student the reason for the use of the word *about* when the results of measurements of lengths and distances are reported? When might the word not be needed?

- Think of and describe examples of when the result of a measurement of a length or distance is rounded to the next higher unit and when it is rounded down to the next lower unit.

- Some of the requests for reports can be revisited after the millimetre is introduced as one-tenth of a centimetre. An optical illusion task, i.e., two perpendicular segments of equal lengths $\perp$ can be used to illustrate to students an example of a need to measure. Sometimes things are not as they appear and our eyes can deceive us.

- A follow-up discussion related to requests that involved measuring the lengths of several objects to the nearest millimetre can be used to reinforce the necessity of using the descriptor *about* when results are recorded and reported. This could lead to the conclusion that all measurement is approximate.

A task can include the request to prepare lists of objects that are:

- About **1 mm** long or thick.
- About **1 cm** long or thick.
- About **1 m** long or tall.

After these lists are compared, a composite list can be prepared and displayed on the bulletin board. The members of the list can serve as *referents* for future estimation tasks.

A discussion about the estimation strategies used by different students to estimate the thickness of a book in millimetres and the width of a book in centimetres should result in the discovery that different estimation strategies exist. Trying a new or different strategy is an aspect of *flexible thinking*.

The ability to relate a unit to a personal *referent* fosters the ability to estimate with that unit.

- Which personal *referent* can students use for a length of *about* **1 cm**?
- What can students use to remind themselves of a length of *about* **1 m**?
- How can students use the information of knowing their heights in centimetres help them to estimate the height of a door or the height of a room in centimetres?

As students work with a partner, they are challenged to find at least two different answers for the following types of questions or requests:

- A capital T and ⊔ made up of two and three segments, respectively are shown on the chalkboard or on a piece of paper. The request is made to think of and to explain how the total length of the segments used to draw each letter could be estimated.

- Describe how you could determine the distance around or the perimeter of a tin can.

- How can you calculate the distance around or the perimeter of a rectangular region that is not a square?

- How can you calculate the distance around or the perimeter of a rectangular region that is a square?

During a follow-up discussion the strategies that were used are compared.

Calculate and record the missing numerals.

6 cm = _ mm **6 m = _ cm** **6 m = _ mm**

What is your rule for changing:

- centimetres to millimetres?
- metres to centimetres?
- metres to millimetres?

Someone measured the length of a piece of string and recorded **2.222 m** as the answer. What is the meaning of each of the twos in the answer?

How would you record each of the following in metres?

4 cm **40 cm** **44 cm** **404 cm**
4 mm **44 mm** **444 mm** **4 444 mm**

Someone made the statement that it is very difficult to measure the length of one's shadow to the nearest millimetre. Why might that be the case?

About Estimation

A student's ability to estimate is related to *number sense*, familiarity with the units that are used and the availability of at least one estimation strategy. This dependence on several important variables implies that at any given time the abilities to estimate will vary greatly among students.

Accommodating Responses:
If as part of estimation tasks, students are asked to look at segments on the chalkboard, it must be kept in mind that an equity issue exists. Students sitting at the back of the room or at the sides will have a different perspective of the segments that are being considered. Permission should be granted to have all students look at the segments from the same viewpoint.

When students volunteer to share their estimates with classmates care is required in acknowledging the responses. It must be assumed that if students have used a strategy to estimate, rather than making a guess, all types of responses require the same type of response. It could be very discouraging for some students if an initial response is labelled as *good*, *very good* or even as *excellent*. Students who arrived at a different numerical value may end up being reluctant to share responses and a loss of confidence could result. An atmosphere is required that lets students know that all estimates are accepted and acknowledged in the same way. However, some responses can be used as diagnostic information which can be dealt with as further estimation experiences become part of an **IEP**.

Assessment Suggestions

The important learning outcomes for measurement of length are related to:

- becoming familiar with the different units of measurement of length.
- being able to estimate lengths using different units as part of a *referent*.
- knowing how to read and use a ruler correctly.

Possible Types of Assessment Tasks

a) Would you use millimetre, centimetre or metre to describe each of the following? Explain your decision.

- The length of a pencil?
- The thickness of a loonie?
- The length of a hallway?

b) Select one of the choices. Explain your thinking.

- What is the length of a truck? **1 m 10 m 100 m**
- What is the height of a pop can? **1 cm 10 cm 100 cm**
- What is the thickness of a book? **2 mm 2 cm 200 cm**

- Which could be used to report the amount of rainfall in one hour?
  8 mm 8 cm 8 m

- Explain, orally or in writing, how you would estimate:
  (i) the length of the gymnasium.
  (ii) the length of one side of a big book.

- Why is the word *about* used when measurements of length are recorded?

- Record the length of each segment or path. Explain your thinking.

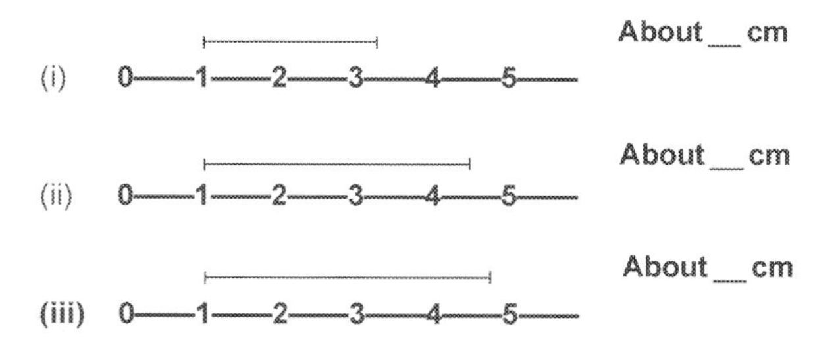

Reporting

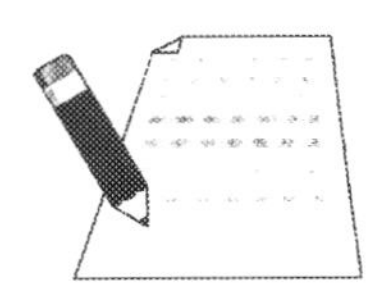

The types of assessment tasks that are suggested can make it possible to share information about:

- Indicators of familiarity with the units of measurement of length; millimetre, centimetre and metre.
- Knowledge of the use of appropriate units for different types of problems that require measurement of length.
- Ability to estimate and to explain the strategy that was employed and the *referent* that was used to arrive at an estimate.
- Knowledge of reading and interpreting the numerals and markings on a ruler.
- Knowledge of recording measured lengths and the reasons for using the term *about* as part of these recordings.

Indicators of *willingness to take risks* and *confidence* can surface when strategies that were used are explained and reasons are given for the responses that were elicited.

For Reflection

What is your reaction to the following scenarios?
What possible purposes do you think the authors of these tasks have had in mind for these types of activities?

1) Instructions for an activity in a reference for students that request estimates to the nearest millimetre.

2) Instructions on a page in a reference for students that ask students to record estimates of the lengths of several segments before measuring the length of each one.

3) Instructions on an activity sheet or on a unit test that ask students to record the estimates for the lengths of several segments.

4) A test about measurement of length that consists of items that ask students to make conversions: from millimetres to centimetres and metres and from centimetres to millimetres and metres.

Measurement of Area – Perimeter

Measurement of area is a new topic for the intermediate grades. The main outcomes for teaching about measurement of area are similar to those for teaching about measurement of length. The sequence of activities and the types of problems described in **Key Ideas from the Previous Grades** (pp.145 and 146) can be adapted for the introduction to and the teaching about measurement of area.[3]

Types of Activities and Problems

- **Definitions – Connecting:** What are examples of statements uttered by a principal, a teacher and someone around the house that use the word *area*? What does the word *area* mean?

  What other word or words could be used instead of *area*?

  How would you explain the meaning of *area* to a young student?

  Who would use the word *perimeter*? When? How?

  What other words could be used instead of *perimeter*?

- **Comparing and Ordering**: Six or more strips of cardboard that are the same widths but different in height are placed randomly on the ledge of the chalkboard.

  What is the same about the strips?

  How do the strips differ?

  What words can you use to describe the differences?

  Guidance may be required to have the students use the terms *size* and *area* rather than the descriptor *big*.

  How could you show someone that any two strips differ in size or area?

  After the strips with the least area and the greatest area are identified, all of the strips are placed in order of area or size. How can the difference between adjacent members in the ordered sequence be described?

  A strip with a different shape is introduced. How can it be determined where this strip would belong in the ordered sequence?

- **Sorting**: A sheet of paper is displayed. The request is made to look at the sheet of paper and to try to identify several objects from around the classroom that are:

  - About the same size or have about the same area.
  - Bigger in size or have more area.
  - Smaller in size or have less area.

  How could it be shown that the suggestions that are made are correct?

- **Measuring Area – Body units:** How might hunters of long ago have measured and described the size or area of the fur of a hunted animal?

  How could the hunter have measured the area?

  What are some possible disadvantages of measuring area or size in this way?

- **Measuring Area – Arbitrary units:** What units could be used in the classroom to measure the area of a small rug or the area of an atlas that would result in answers that would be the very close to being the same for everyone in the classroom?

  What strategy would you use to estimate the size or area of the rug and atlas?
  What would you do to calculate the areas?
  Why is the word *about* needed as part of the answers?

- **Our Eyes Can Deceive Us**: Each student or group of students is given six identical rectangular shapes. The task consists of cutting each of the shapes once and then rearranging and tracing around the pieces. Each cut has to be different from the preceding cut.

  What is different about the traced figures? What is the same?
  If these shapes were pieces of cake, which shape
  do you think a young child would choose?
  Why would that be the case?

- **Standard Units – Referents for Estimating**: With a partner generate a list for: Who is or might be interested in measuring area? When? Where? Why?

  After the need for a standard unit has been established, the square centimetre is introduced.

  Which of your fingernails has an area of about one square centimetre?
  How could you use this information to estimate the area
  of the wrapper for a piece of gum?

  Since square centimetres are too small to handle to solve area measurement tasks, figures are traced onto a grid of square centimetres or are placed on a transparent grid.

  The request is made to outline on grid paper different looking gardens that have areas of six square centimetres with the proviso that each square must have at least one side in common with another square.

  How are the gardens different? How are the same?
  Why might a young child think that the areas of the gardens are not the same?

  Explain one possible strategy for estimating the area of your hand.
  Measure that area of your hand to the nearest ten square centimetres.

  Estimate the area of several different books. Explain your thinking.

  After the need for the square metre has been established its area is illustrated on the floor. Do you think there is anything in the classroom that has an area of about one square metre?

  Be ready to explain the strategy that was used to determine answers for,

  About how many students could sit inside the area of one square metre?
  About how many mathematics books could fit into one square metre?

- **Formula for Area of Rectangular Regions:** During the introduction of the lesson, Mr. C. guided his students to realize that there are advantages to labelleling the dimensions of rectangular regions *base* and *height* rather than *length* and *width*. The students also were reminded that *by* can be used to describe rectangular regions – *two by seven*.

The students were provided with a grid of square centimetres and they were challenged to outline as many different rectangular regions with an area of twenty four square centimetres as they could think of.

The results of the sketches were summarized on the chalkboard in a chart with headings: **Base** **Height** **Area**. As responses were volunteered and recorded, the spacing in the chart was organized to allow recordings to be in order from **24 by 1 = 24** to **1 by 24 = 24**. This recording procedure enabled students to identify the one possibility that they had missed, **8 by 3**.

The students were requested to talk to a partner as they examined the recordings and to come up with a rule for calculating the area for rectangular regions. After listening to the groups during their discussions, Mr. C. announced that he had heard three different rules and that they all were correct. The rules these students generated were:

Count the number of squares along the height and skip count along the base.
Count the number of squares along the base and multiply by the number of squares along the height.
Multiply the number from the base times the number from the height.

These rules were then summarized as:
Area is equal to base times height or **A = b x h**.

The setting illustrates teaching *through problem solving* with the positive outcome that the students know that the **b** and the **h** are representative of number of squares. It was explained that **A** = **l x w** has the same meaning.

- **Perimeter and Area:** The request is made to sketch on a square centimetre grid as many different rectangular regions as possible for a given perimeter, i.e., **16 cm**.
The results of the sketches are to be listed in a table with the headings:
Perimeter 16 cm **Base** **Height** **Area**

As students talk to a partner, they are requested look at the entries in the table and to record at least one statement about rectangular shapes that have the same perimeter. The statements are compared.

  - The request is made to work with a partner and to prepare a reaction and response for the following scenario,
    Why did the foot shrink?
    A student was asked to calculate the area of his foot. The student put a string around his foot, made a shape of a square with it and then multiplied **4 units** times **4 units** to get an answer of **16 square units**. A friend told the student that the foot is shaped like a rectangle. The string was re-shaped into **3** *by* **5** rectangle with an area of **15 square units**.
    What happened to the missing square unit?

  - The request is made to prepare a response for,
    What would you say to someone who suggests that a longer perimeter always means a rectangle with a greater area?

Assessment Suggestions

Key ideas for measurement of area include:

- There are times when things may be different than they appear and there exists a need to measure area.
- The use of personal *referents* as part of estimation strategies involving square centimetres and square metres.
- The ability to explain how the area of rectangular regions can be calculated.
- The ability to explain the relationships between area and perimeter: rectangular regions with the same perimeter can have different areas and rectangular regions with the same area can have different perimeters.

Possible Types of Assessment Tasks

To probe understanding oral or written responses could be selected for requests of the following type:

- Suggest why someone might be interested in calculating the area of a rectangular region.
- Suggest why someone might be interested in measuring the perimeter of a rectangular region.
- Explain two different ways of calculating the perimeter of a rectangular region.
- Why is it not always possible to tell which of several figures has the least or the greatest area?
- Explain how you would estimate the area of a small rectangular region like an envelope in square centimetres.
- Explain how you would estimate in square metres the area of a wall in a room.
- What would you say to someone who states that all figures that have the same area look the same?
- What would you say to someone who states that all figures that have the same perimeter have the same area?
- What would you say to someone who states that all figures that have the same area have the same perimeter?
- The formula for finding the area of a rectangle is **A = b x h** or **A = l x w**. Explain the meaning of each of the letters and why **b** and **h** are multiplied to calculate the area of a rectangular region.

Reporting

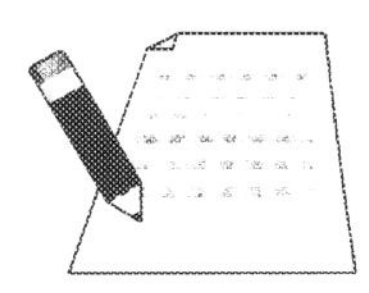

These types of assessment tasks make it possible to share information about:

- Ability to *connect* area and perimeter to settings or actions outside the classroom.
- Ability to explain strategies for estimating areas of rectangular regions in terms of the *referents* that are used.
- Ability to explain the derivation of the formula and the meaning of each part of the formula for calculating the area of rectangular regions.
- Ability to illustrate that rectangular regions with the same area or the same perimeter can look different
- Ability to illustrate that rectangular regions with perimeters that differ can have the same area.

For Reflection

What is your reaction to the following types of assessment tasks?

1) As part of an end-of-unit test about measurement of area students are asked to calculate the area of five rectangular regions.

2) As part of an end-of-unit test about measurement of area students are asked to calculate the area of five rectangular regions. Different units, centimetres and millimetres, are used for the dimensions for two of the rectangular regions.

3) As part of an end-of-unit test about measurement of area and perimeter students are asked to calculate the perimeter of five different figures.

Measurement of Capacity and Volume

Capacity

The results of measuring the capacity of containers provide answers to the questions:
- How much liquid does a partially filled container hold?
- How much liquid can a full container hold?
- Which container holds more liquid?

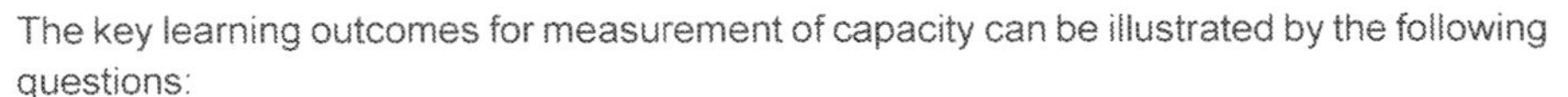

The key learning outcomes for measurement of capacity can be illustrated by the following questions:
- Who talks about liquid capacity of containers? When? Where? Why?
- Who measures capacity? Where? Why?
- Why is it necessary to measure the capacity of containers?
- How can the capacity of containers be measured?
- What are the standard units for measuring the capacity of containers?
- What is the relationship between the standards units for measuring capacity?
- What are some of the challenges that are part of estimating the capacity of containers and what are possible *referents* for estimating capacity?

The sequence of presenting the main ideas can be very similar to the one used for measurement of area.

Types of Activities and Problems

- **Connecting – Litres:** Several one litre containers of different shapes are displayed.

  What is different about the containers?
  Try to think of something that is and could be the same for all of the containers.
  What do you think it is?

  Some guidance may be required. The questions could be asked,

  Why might some young students believe that some of the containers can hold different amounts of liquid?

  After it is determined that when these different looking containers are filled they all hold the same amount of liquid, **one litre**, the students could be invited to make suggestions for,

  Why do you think a capital **L** *rather than a small* **l** *is used as the symbol for litre or litres?*

  A request is made to generate a list of things or items that are sold in litres. The lists are compared. A composite list is prepared for display purposes.

- **Estimating – Litres:** An ice cream bucket is displayed.

  *Which of the **1L** containers would you use to estimate the capacity of the ice cream bucket?*
  Explain your thinking.

Different containers are displayed.
For example,

watering can   large pail   plastic container

The requests are made to,

Estimate the capacity of the containers.
Use the word about to record the estimates, ***About _ L****, and*
be ready to explain your estimation strategy.

The introduction of the one thousand base ten block can foster *visualization.* Some students may prefer to use this block as a *referent* for estimating the capacity of containers in litres. As part of the discussion, students are told that a container with the dimensions of this block, **10 cm** by **10 cm** by **10 cm** can hold **one litre** of water.

As more estimation tasks are presented some students may use known capacities for certain containers as a *referent,* like the **4 L** ice cream bucket.

Explain your strategy for estimating the capacity of:

- *an aquarium.*
- *a sink.*
- *a bathtub.*

Accommodating Responses:
Estimating capacity of containers is a challenging task since three dimensions need to be considered. It must be kept in mind that for extreme differences in dimensions some students' comparisons and estimates can be influenced by what they perceive. Their eyes may tell and still convince them that a very tall and narrow container must be able to hold more liquid than a very short and wide container.

- **Connecting and Thinking about Millilitres:** A **1 mL** measuring spoon, an eyedropper and the dimensions of a cubic centimetre can be used to introduce one one-thousandth or $\frac{1}{1000}$ or **0.001** of a litre.

The request is made to generate a list of items or things that are packaged and sold in millilitres. The lists are compared and a composite list is displayed.

Entries from the displayed list can be selected and presented along with several choices for answers. Familiarity with decimals can determine the format of the choices, as illustrated by the 'Tube of toothpaste' example below.

Requests of the following type are made:

Which choice obviously does not make sense to you? Why is that the case?
Which choice or choices make sense to you?
Explain your thinking.

For example,

- Tube of toothpaste: **13 mL** **130 mL** **1300 mL**
  Or **1.3 mL** **13.0 mL** **130.0 mL**
- Can of juice: **350 mL** **3500 mL** **35 mL**
- Bottle of pop: **20 mL** **2 mL** **200 mL**
- Shampoo bottle: **3 mL** **37 mL** **375 mL**
- Amount of macaroni for 4 servings: **5 mL** **50 mL** **500 mL**

Pretend you are baking a chocolate cake of 9 servings.
Which amounts would you choose for the recipe? Be ready to explain your thinking.
Compare your selections with those made by others.

- Salt – **2 mL** **20 mL** **200 mL**
- Milk – **25 mL** **250 mL** **2500 mL**
- Oil – **7.5 mL** **75 mL** **750 mL**
- Sugar – **2.5 mL** **25 mL** **250 mL**
- Ground cinnamon – **2 mL** **20 mL** **200 mL**

Bake your cake. Enjoy!

Volume

The questions that are posed about capacity and measurement of capacity are relevant for tasks related to measurement of volume. A teaching sequence for measurement of volume can be developed by examining the main ideas for the sequence suggested for measurement of area.

Types of Activities and Problems

- **Definition – Connecting:** During an introduction the different meanings and uses of the word *volume* are discussed. Two rectangular prisms that differ in size are shown as students are guided to focus on the fact that the blocks differ in size and they take up different amounts of space. Two empty boxes both in the shape of rectangular prisms but different in size are compared.

  How are the boxes different? How can the differences be described?

  As part of the discussion students are guided to describe the differences by using terms like:

  - more room or more space inside or a larger volume or,
  - less room, less space inside or smaller volume.

  At the conclusion of the discussion the following requests are made,

  What other words could be used to describe the volume of a box to someone?
  Who might be interested in talking about volume of boxes?
  Who might be interested in comparing and measuring the volume of boxes?

- **Comparing and Ordering:** A collection of boxes of similar shapes but different in size is examined.

  What is the same about the boxes?
  How do the boxes differ?

  As one box is held up, the requests are made to identify one box that has a larger volume and one that has a smaller volume.

  How can it be shown that the selections are correct?

  The boxes are put in order of volume. Students are requested to describe, orally or in writing, the differences between two adjacent members in the ordered sequence.

  Who might be interested in considering volume and putting boxes in order?

- **Our Eyes Can Deceive Us:** From a collection of blocks, sets of twelve blocks of the same size are displayed on a table. The students are requested to think of a rectangular solid that can be constructed from a set of twelve blocks. Following each example that has been erected on the table, the request is made to think of an example that looks different from those on the table.
At the conclusion, the buildings are examined.

  How are the buildings different?
  What are all of the things that are the same for all of the buildings?

  Guidance may be required to think of each of the blocks as a room of the buildings.

  Why might a young student think that one of the buildings has more room or has more volume?
  Which building might a young child choose as the one with the most room or volume?
  Why might that be the case?

  The conclusion that things may look or appear to be different but can be or are the same as far as an important characteristic is concerned should be part of the discussion. A further illustration of this can be the presentation of two rectangular boxes whose appearances differ but have the same volumes. The discussion should conclude with students seeing a need for units to measure volume.

- **Measuring Volume – Body Units and Arbitrary Units:** It is possible to describe the volume of a rectangular box by using something like hands full of sand as a unit of measurement.

  What are possible disadvantages of using hands full of sand to measure and describe the volume of a box?

  Boxes could be filled with marbles and a marble count could be used to describe the volume of boxes.

  What are possible disadvantages of using marbles to measure and describe the volume of boxes?

  Sugar cubes or blocks of any type can be used to fill and to describe the volume of boxes.

  What are possible disadvantages of using sugar cubes or any kind of block to measure and describe the volume of boxes?

  The goal of this type of an activity setting is to have students see the need for a standard unit.

- **Measuring Volume – Standard Units:** The cubic centimetre and cubic metre are introduced.

  A cube with dimensions **1 cm** by **1 cm** by **1 cm** is called a cubic centimetre.

  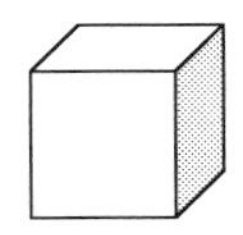

  Which tip of your fingers is about one cubic centimetre?
  How many of these fingertips do you think would fit into this match box?

  The fingertip selected by students could become a *referent* for estimating the volume of small rectangular boxes.

  How would or could you estimate the volume of a small crayon box?
  Explain your strategy.

  The strategies that the students suggest are compared.

Twelve sticks one metre long are used to build a model of a cubic metre. Students are requested to be ready to explain the strategies used to arrive at the answers for:

About how many recycling boxes do you think would fit into one cubic metre?
How many students just like you do you think could fit into one cubic metre?
How would you estimate the volume of this classroom in cubic metres?

The strategies are compared during follow-up discussions.

The students are requested to prepare lists for:

What are some things that are measured in cubic centimetres?
What are some things that are measured in cubic metres?

The lists are compared. A composite list is displayed on the bulletin board.

- **Formula for Volume:** As different rectangular prisms are constructed from blocks, the number of blocks along the length, the width and the height are counted and recorded in a chart. The blocks for the prisms or buildings are counted and the volume is recorded in the chart.
For example,

Number of Blocks			
Length	**Width**	**Height**	**Volume**
3	2	2	12

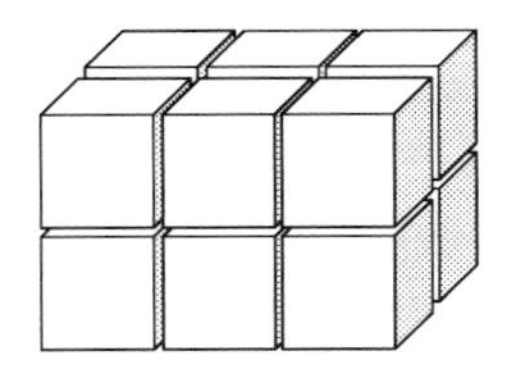

After several examples, the challenge is presented to,

Try and tell how volume can be calculated from the entries for length, width and height.
What would a rule be for calculating the volume of rectangular solids?

The dimensions of a rectangular box are measured to the nearest centimetre and the results are recorded in a chart.
For example,

Number of Centimetres		
Length	**Width**	**Height**
About ___ cm	**About ___ cm**	**About ___ cm**

How does knowing the length and the width help in finding out about how many cubic centimetres there are in one layer or in the bottom layer of a box?
How do you know about how many layers there are in a box?
What is a rule for finding the volume of a box if the length, the width and the height are known?
Why is the word about used when the volume of a box is recorded?

- **Designing Boxes:** The students are requested to cut three **12 cm** by **10 cm** rectangles from centimetre grid paper.
For one rectangle, squares of side **1 cm** are cut from each corner. For the other two rectangles squares of side **2 cm** and **3 cm** are cut from each corner, respectively. The edges of each piece are folded to make an open box.

The dimensions and the volume for each box are recorded in a chart:

Length	Width	Height	Volume
___ cm	___ cm	___ cm	___ cm^3
___ cm	___ cm	___ cm	___ cm^3
___ cm	___ cm	___ cm	___ cm^3

Use the entries from the chart and write a sentence about boxes and volume.
Who might be interested in designing boxes?
What do you think you could do to find out whether or not it is possible to design a box with greater volume than the box with the greatest volume in the chart?

Assessment Suggestions

Some of the key ideas related to measurement of capacity and volume include:

- Who is interested in the measurement of capacity and volume?
- When might the measurement of capacity and volume be required or necessary?
- What standard units are used to measure capacity and volume and what is the relationship between these units?
- Familiarity with the units used to measure capacity and volume.
- What is the formula for calculating the volume of rectangular prisms and what does each part of the formula mean?
- What are some possible strategies for estimating capacity and volume?
- Why can estimating capacity and volume be a challenge?

Possible Types of Assessment Tasks

a) State what you think you need to know for each situation:
Area, **Perimeter**, **Capacity** or **Volume**.

- How many posters fit on the bulletin board?
- How many boxes of popcorn can be filled from this batch?
- The box needs a ribbon around it.
- The knee of the jeans requires a patch.
- I need a bigger pot to reheat the soup.
- I think your cup is bigger than my cup.
- I think my hand is bigger than your hand.
- I think your wrist is bigger than my wrist.
- How many books will fit into this box?
- Does this jug hold enough lemonade for all of us?

b) Which unit would you use for each measurement? Explain your thinking.

- The area of a book: **square centimetres** **square metres**
- The capacity of an aquarium: **millilitres** **litres**
- The volume of a pencil box: **cubic centimetres** **cubic metres**
- The size of a carpet: **square centimetres** **square metres**

continued next page

c) Write your own definition for each:
(i) Area **(ii) Perimeter** **(iii) Capacity** **(iv) Volume**

d) Name at least one type of work that involves the measurement of:
(i) Area **(ii) Perimeter** **(iii) Capacity** **(iv) Volume**

e) Select an answer. Explain your thinking.
- capacity of a tube of toothpaste: **2 mL** **200 mL**
- capacity of an aquarium: **2 L** **20 L**
- area of a mathematics book: **45 cm^2** **450 cm^2**
- volume of a box in the shape and size of a mathematics book: **100 cm^3** **1000 cm^3**
- perimeter of a mathematics book: **90 cm** **900 cm**

f) Explain how the volume of a rectangular prism or box can be calculated.
- What is the formula for calculating the volume of rectangular boxes?
- What does each part of the formula mean and why are the parts multiplied?
- Why is the word *about* used when the result of a measurement of volume is recorded?
- What are some possible reasons for measuring the volume of boxes?

g) Explain your estimation strategy for:
- About how many millilitres can a paper cup hold?
- About how many litres can a pail hold?
- About how many cubic centimetres in a small crayon box?

h) What would you say to someone who states that all boxes that look different have different volumes?

i) What would you say to someone who states that all containers that look different hold different amounts of liquid when they are filled?

Reporting

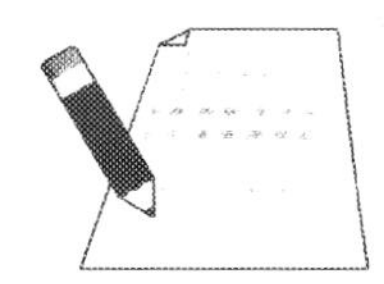

These types of assessment tasks make it possible to share information about:

- Ability to *connect* notions about area, perimeter, capacity and volume to events and actions outside the classroom.
- Understanding of the terms area, perimeter, capacity and volume since the request is made to define these terms in their own words.
- Familiarity with the units of measurement for capacity and volume.
- Ability to explain the derivation of the formula for calculating the volume of rectangular prisms and the meaning of the terms in the formula.
- Ability to explain estimation strategies for millilitres, litres and cubic centimetres.
- Realization that as far as capacity and volume is concerned containers and rectangular prisms may be different from what our eyes may lead us to believe.

For Reflection

1) How would you use procedures and ideas related to measurement of area, capacity and volume to explain to someone the meaning of *measurement sense*?

2) What examples from areas of measurement would you use to illustrate an explanation of the impact of an emphasis on *conceptual understanding* and *cognitive processes* on assessment procedures?

Chapter 7 – Geometry and Spatial Sense

According to the curriculum[1] *spatial sense* involves *visualization*, *mental imagery* and *spatial reasoning* and it offers a way to interpret and reflect on the physical environment (p.12). *Spatial sense* is essential in many tasks such as reading tables for information, making diagrams, reading maps and visualizing objects that are described orally.[2] *Spatial sense* is required for the construction and interpretation of graphs and charts. The interpretation of game boards or of personal spaces requires aspects of *spatial sense*.

Spatial sense can be developed and enhanced through appropriately planned instructional settings. As is the case for *number sense* and *measurement sense*, students require many experiences that provide opportunities to explore properties and relationships as well as interrelationships of three dimensional and two dimensional figures. Opportunities need to be provided to share the results of explorations orally as well as in writing.

The lessons that are planned and presented need to accommodate the levels of development of geometric understanding or geometric thinking developed by two Dutch educators, Dina van Hiele-Geldorf and Pierre van Hiele.[3]

- **Level 0 – Visualization:** Students recognize and name figures by their global appearances in common positions.
- **Level 1 – Analysis:** Students observe the component parts of figures without explaining possible relationships that may exist.
- **Level 2 – Informal Deduction:** Students can deduce properties of figures and express relationships both within and between figures.
- **Level 3 – Formal Deduction:** Students at this level are able to work with abstract statements about geometric properties and draw conclusions based more on logic than intuition.
- **Level 4 – Rigour:** This level includes the ability to analyze and compare different axiomatic systems in a rigorous way, an ability used by geometricians and mathematicians.

According to the van Hieles, instruction and experience rather than age or maturation play an important role in students' progress through the sequential levels. Functioning successfully at one level requires the capabilities of the preceding level. Lessons that deal with specific aspects of geometry and *spatial sense* should consider the thought levels of the students. During lessons, opportunities should be provided that challenge students to move to the next thought level. It is the thinking that is the valuable part of any lesson, rather than any specific content.

Key Ideas from the Previous Grades

The learning outcomes related to geometric thinking in the primary grades focused on *visualization* and analysis.

Types of Activities and Problems for three-dimensional figures included:

- Sorting 3-D figures by relating them to objects encountered in the environment. Reaching the conclusion that a different viewpoint of a figure may result in connections to different objects in the environment.

- Faces (edges; vertices) on 3-D figures were identified and compared. The figures were sorted according to the number of faces (edges; vertices).

- Pictorial representations of 3-D figures were matched with the appropriate figures. The figures were held as indicated in the representations. Parts that were visible and not visible in the representations were identified.

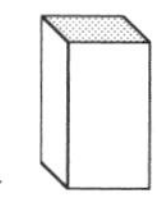

- The number of faces, edges and vertices that could be seen in pictorial representations of 3-D figures were counted. Students speculated about the possible number of faces, edges and vertices that could not be seen in the representation.

- For a display of four different 3-D figures and the question,
  Which do you think does not belong? or,
  Which of these do you think is different?
  at least two 3-D figures were identified by making references to number and/or shape of faces, to number and/or shape of edges or to number of vertices.

- At least two differences were identified for 3-D figures placed next to each other in a linear sequence.

- Students speculated about parts of drawings, photographs, or shadows on an overhead of a 3-D figure that were not visible.

Types of Activities and Problems for two-dimensional figures included:

- The number of sides and corners for special shapes were identified: rectangles, triangles, circles, pentagons, hexagons and octagons.

- General terms were used to describe how special shapes of the same type can differ.

Since *spatial sense* is part of the conceptual domain and it develops over time, many of the key ideas from the previous grades may have to be revisited, perhaps even more than once for some students.

Polyhedra – Prisms

Three dimensional figures can be sorted in different ways. One possible subset of these figures is the set of solids which are bounded by plane surfaces only. These solids are called polyhedra. Two types of polyhedra are introduced, *rectangular prisms* and *triangular prisms*.

Types of Activities and Problems

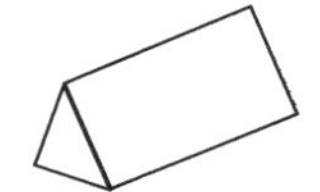

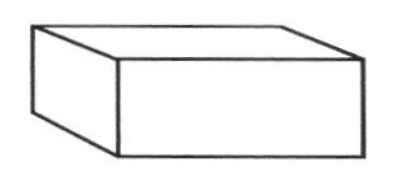

- **Definitions**
  Two sets of polyhedra are presented. Set A consists of *rectangular prisms* of different sizes and set B of *triangular prisms* of different sizes. A reminder may be needed for the fact that *cubes* are special *rectangular prisms*.

Some of the activities and problems that are described can be presented twice, once without cubes and then with cubes as part of the collection of *rectangular prisms*.

As students examine the two sets they are asked to respond orally or in writing to requests of the following type:

- *Describe how the solids in each set are the same and how they differ.*

- The blocks in both sets are called *prisms*.
  What do you think the word prism means or what do you think it could mean? Explain your thinking.

- The prisms in set A are called *rectangular prisms*.
  Try to think of a reason for assigning this name to these prisms.
  The prisms in set B are called *triangular prisms*.
  Try to think of a reason for assigning this name to these prisms.

- *With the aid of a sketch how would you explain the meanings of rectangular prism and triangular prism to a young student?*
  How would you do it without a sketch?

- **Visualizing Attributes**
  - A *rectangular prism* is held up in a stationary position at the front of the classroom. The request is made to show with fingers on one hand how many faces of the *prism* can be seen and with the fingers on the other hand how many faces students think they cannot see.
    Why are the responses different?
    How can you tell by looking at your fingers that your response is correct?

  - A sketch or a photograph of a *rectangular* and a *triangular prism* is presented. The request is made to record answers for:
    How many faces, edges and vertices can be seen?
    How many faces, edges and vertices cannot be seen?
    How do you know your recorded answers are correct?

  - The following scenario is presented:
    If you know that you are looking at a shadow of a 3-D figure, what do you think that figure could look like if the shadow:
    - is a rectangular region that is not a square?
    - is a rectangular region that is a square?
    - is a triangular region?
    Explain your thinking by selecting and showing the 3-D figures you have in mind.

- **Faces and Nets**
  A *sketch of a net for a tetrahedron could be used to explain the meaning of the term.*
  - The challenge is presented of drawing a net for a *rectangular prism* and *triangular prism* on pieces of cardboard.
    After the drawings are compared the request is made to try and fold the cardboard to show the prisms and to,
    Record a definition of net that you could share with a younger student.
    The definitions are compared.

- Several different nets are presented and the request is made to,
  Try and predict which nets can be folded into rectangular prisms or triangular prisms and which cannot.
  How was the decision made? Explain your thinking.

- **Constructions – Pattern – Instructions**
  - After models of a rectangular and a triangular prism are constructed from Plasticine, the request is made to,
    - *Use sticks or toothpicks and small balls of Plasticine (pieces of straw and pipe cleaners) to create models of the two types of prisms.*
    - *Count and record the number of edges, vertices and faces for each prism.*
    - *Given the entries for faces and vertices, how is it possible to predict the entry for the number of edges?*
    - *Check to see if your prediction works for two other blocks? What did you find out? Record your observation.*

  - Written instructions for a classmate.
    - *Think of and describe attributes that will lead a classmate to select from a collection of 3-D figures:*
      - *A triangular prism.*
      - *A rectangular prism that is not a cube.*
      - *A rectangular prism that is a cube.*

    - *Compare your instructions with those created by a classmate. How are they the same? How are they different?*

      With a partner try to record the least number of hints you think you could give to someone for selecting a rectangular prism that is not a cube and for selecting a triangular prism from a collection of 3-D objects. Compare your recordings.

    - *Think of and describe attributes that will lead a classmate to construct with sticks and Plasticine a skeleton of:*
      - *A triangular prism.*
      - *A rectangular prism that is not a cube.*
      - *A rectangular prism that is a cube.*

- **Connecting**
  Requests of the following type are presented:
  - *Prepare a list of objects that are shaped like rectangular prisms.*
    Who might be interested in these shapes?
    Compare your list and suggestions.

  - *Prepare lists of objects that are shaped like triangular prisms.*
    Who might be interested in these shapes?
    Compare your list and suggestions.

Angles, Measurement of Angles and Triangles

The ability to define angle, describe how angles are measured and the classification and naming of selected angles are pre-requisites for activities and problems that involve the levels of geometric understanding labelled analysis and informal deduction.

Angles can be defined by making reference to acts of turning and the familiar language that is related to experiences outside the classroom– *left turn*; *right turn*; *about turn*; and *turn all the way around*.

An angle can be thought of as a turn or as the shape of two segments with a common endpoint. The hands of a clock, the arms of a protractor or a person, the blades of a pair of scissors and the thumb separating from the remaining fingers can be used to illustrate angles. The use of a person's arms and the arms of a protractor illustrates the important idea that the shape or opening called angle is independent of the lengths of the arms or segments.

Introductory Activities

- Early activities with angles can include:
  - Comparing angles:

    Show how you know that one angle is greater than or less than another angle.
  - Sorting angles into different categories:

    For example, angles whose measures are greater than or less than the measure of a given angle or are about the same.
  - Ordering angles:

    Tracing angles and then matching arms and common endpoints to order the measures of the angles from least to greatest or vice versa.

A right turn or a quarter turn is defined as a *right* angle.

Accommodating Responses

There may be some students who will be tempted to call a quarter turn to the left a *left* angle. This is a logical deduction, but a reminder will be required that this is not the case.

- The *right* angle can be used as a *referent* for the preparation of a list along with sketches of parts of objects in the classroom that have *right* angles. The lists are compared. A composite list is prepared for a display.

- The *right angle* can be used as a *referent* for the preparation of a list of parts of objects that are examples of angles with measures less than right angles. The lists are compared and a composite list with an appropriate heading is displayed. Students are told that angles that have a measure less than a right angle are identified as *acute* angles.

  Write a sentence about how you will try to remember the meaning of acute angle.

  The answers are compared.

- Two *right* angles placed side by side are defined as a *straight* angle.

  Write a sentence to explain why you think two right angles are called a straight angle.

- The *right* angle can be used as a *referent* for the preparation of a list of parts of objects that are examples of angles that are greater than right angles. The lists are compared and a composite list with an appropriate heading is displayed. After being told that angles that have a measure greater than a right angle and less than a straight angle are identified as *obtuse* angles, the request is made to,

  Write a sentence about how you will try to remember the meaning of obtuse angle.

  The sentences are compared.

- The request is made to write a paragraph to answer the following questions,

  Who might be interested in talking and thinking about angles?
  Who might care whether or not the measures of angles are acute or obtuse?
  What might be a good title for the paragraph that includes the word angle or angles?

  The paragraphs are shared and compared.

There are times when it is important to know the answers for the questions,

How big is an angle?
How far is the turn between the two arms?

The answer is reported in *degrees*, the same word that is used for reporting temperature. A complete or *full turn* is divided into **360** equal parts or **360** degrees or **360°**. That means that a quarter turn or a *right* angle is **90°** and one-half of a turn or a *straight* angle is **180°**.

Why **360** and where did **360** originate? A long time ago the Babylonians did observe that the position of the sun changed by the same amount each day of the year and they called this change one degree. Since number of days in a year or **365** had very few divisors, the Babylonians decided to divide the circle into **360** equal parts or **360 °** instead.[(4)]

Examples as well as non-examples can be used to illustrate the points that need to be kept in mind as a protractor is used to record angle measurements. Demonstrations need to consider the correct placement of the baseline and the vertex.

For activities that involve estimating the size or measures of angles, one-half of a right angle or **45°** can be used as a *referent*. When estimation tasks include the examination of angles sketched on the chalkboard, every student in the room needs to have the same viewpoint.

Measuring and Estimating Angles – Types of Triangles

Types of Activities and Problems

- **Definition of Angle**
  Write a note to someone who thinks that the longer the arms or segments of an angle, the greater the measure or size of the angle.

- **Measurement of Angles**
  Write instructions for someone that explain how to use a protractor to measure an angle. Include comments about what should not be done and why that is the case.

- **Definitions**
  Write a note to explain to someone the meanings of acute angle and obtuse angle. Tell why you will never forget the meanings of acute and obtuse.

- **Estimating Measures of Acute Angles**
  Write a note to explain to someone your strategy for estimating the measure of acute angles that are shown in different positions.

- **Estimating Measures of Obtuse Angles**
  Write a note to explain to someone your strategy for estimating the measure of an obtuse angle.

- **Sorting and Naming Triangular Shapes – Sum of the Angles**
  A collection of triangular shapes that includes several examples of each of the six different types of triangles is presented to pairs of students.
  The request is made to,
  Use one angle or angles that are the same to sort the shapes. After the sorting is completed discuss each of the following questions and be ready to share your responses or conclusions.

  - *If one group of the sorted shapes is labelled obtuse-angled triangles, which one do you think it would be? Why do you think so?*

  - *If one group of the sorted shapes is labelled acute-angled triangles, which one do you think it would be? Why do you think so?*

  - *If one group of the sorted shapes is labelled right-angled triangles, which one do you think it would be? Why do you think so?*

  - *Look up the meaning of equilateral. Make a sketch of an equilateral triangle. Write a note to someone to explain how to recognize an equilateral triangle and how you think the meaning of the word equilateral may be remembered.*
    Compare your note to several notes created by classmates.

  After triangles with two equal sides are identified as *isosceles* triangles and those with no sides equal as *scalene* triangles, the request is made to,

  - *Try and think of possible ways of remembering the meanings of isosceles and scalene.*

    If deemed necessary, hints could be given to look at parts of the words or to pronounce the words in different ways. The strategies that students generate are shared.

  - The request is made to shuffle the sorted triangles. Turns are taken. One student selects a triangle and the partner attempts to say something about an angle or about the angles of the triangle and tries to name the type of triangle. As the task is repeated, the triangles are to be held in different positions.

  - Requests of the following type are presented,
    Tear off the three angles of any triangle and push the points or vertices together.
    What did you find out?
    What can you say about the sum of the angles of the triangle?
    Do you think the sum is the same or is different for another triangle?
    Explain your thinking.
    What do you think would be the result if you tear off the four angles of a rectangle and push these together?
    Check to see if you are right.

Parts of 3-Dimensional and 2-Dimensional Figures

Types of Activities and Problems

- **Definitions – Connecting**
  The following terms are displayed on a chart or on the chalkboard:
  parallel vertical horizontal perpendicular intersecting
  As an example from around the room is depicted, the request is made to try to identify the term or terms that can be used to describe the example.
  - The ledge of the window.
  - The long sides of the door.
  - The long side and the short side of the chalkboard.
  - The front wall and the side wall of the room.
  - The two side walls of the room.
  - The two traced diagonals of a folded piece of paper.

  After at least two examples have been described for each term, the request is made to,
  Write a note to someone to explain the meaning of each term.
  Tell how you think the meaning of each term might be remembered.

  These definitions and hints to remember the meanings are shared. The definitions could also be compared to those listed in a dictionary.

- **Parts of Solids – Riddles**
  Several different solids are provided to pairs of students. As statements are made that include each of the following descriptors, the request is made to point to the edges or faces that illustrate the meaning of the descriptors in these statements.
  - *Parallel* edges.
  - *Edges not parallel.*
  - *Perpendicular faces.*
  - *Faces not perpendicular.*
  - *Vertical* edges.
  - *Edges not vertical.*
  - *Horizontal* face.
  - *Face not horizontal.*
  - *Edges meet at a right* angle.
  - *Edges meet at an acute* angle.
  - *Face in the shape of a right-angled* triangle.
  - *Face in the shape of an isosceles* triangle.

  The students are requested to take turns in describing a part or parts of a solid until the partner can identify the solid. A different descriptor is to be used for each turn. For example,
  The solid does not have any faces that are parallel.
  Two pairs of edges are parallel.

  The solid has two faces that are parallel.
  Two faces are perpendicular to each other.
  Four pairs of edges are parallel.

- **Sketching 2-D Figures – Following Instructions**

  The challenge is presented of trying to prepare sketches from pieces of information that are given, one at a time. The goal is to identify the shape of the figure.

  The students could work with a partner. After each piece of information about a figure, statements could be made about what the figure could not be and guesses could be made about what the figure could be.

  What does the figure look like?

  - A figure with four sides. – Two pairs of parallel sides.
    – The pairs of sides differ in length. – There are no right angles.
  - A figure with four sides. – Two pairs of parallel sides.
    – The pairs of sides differ in length. – There are four right angles.
  - A figure with four sides. – Two pairs of parallel sides.
    – The sides are equal in length. – There are four right angles.
  - A figure with four sides. – Two pairs of parallel sides.
    – The sides are equal in length. – There are no right angles.
  - A figure with four sides. – One pair of sides is parallel.
    – The parallel sides are not equal in length. – There is one right angle.
  - A figure with four sides. – One pair of sides is parallel.
    – The parallel sides differ in length. – There are no right angles.
    – The sides that are not parallel are equal in length.

The following terms are displayed:

trapezoid *parallelogram* *rectangle* *square* *rhombus*

The students are told that the figures with the special shapes that were sketched from the instructions that were given can be identified by these names.

Which name do you think goes with which shape or shapes?
Discuss this with your partner and explain your reasons for choosing the name or making a guess about it.

If necessary, the correct labels are presented after the responses have been compared.

Write a note to someone to explain how you try to remember the meaning of each label for these special shapes.

The strategies for remembering the terms are compared.

The request is made to,
Prepare a sketch of a figure with six sides. Record instructions which would make it possible for someone to draw the sketch by following these instructions.
Present the instruction to someone. Were the sketches the same?

How many instructions did it take?
Do you think you could have had someone draw the appropriate figure using fewer instructions? Why or why not?

Transformations and Symmetry

The ability to predict and to explain the results of movements or *transformations* of objects or figures is one aspect of *spatial reasoning* or *spatial sense*.

- **Translation**
  Sometimes objects are moved to-and-fro or up-and-down in a straight line without any turning taking place. The location is changed but the shape and the size remains the same. This type of movement is called a *translation* or a *slide*.

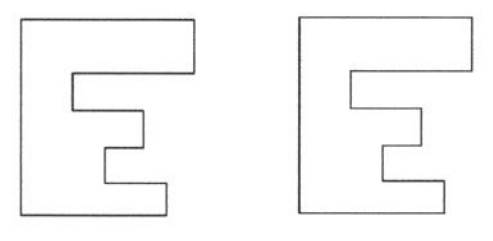

Translations or *slides* can be illustrated by pushing a chair from one place to another or by pushing a solid or book on a table from one position to another in a straight line. An attempt to illustrate a slide could include trying to change a student's position while the student remains as rigid as possible.

A chair at the front of the room is pushed away from the students in one direction. The information is shared that the front legs of the chair or the legs nearest the students moved two metres.

How far do you think the back legs of the chair moved?

After it is concluded that all parts of the chair moved the same distance in a straight line, this movement is labelled a *slide* or a *translation*.

What might be a way to remember the meaning of slide and translation?

- Centimetre grid paper is provided.
  The following requests are made,
  Print a capital A somewhere near the middle of the grid paper with the point and the ends of the 'legs' located on intersections of the grid.
  a) *Slide the A to the right.*
  b) *Slide the A to the upper right.*
  c) *Slide the A to the lower left.*
  Write a sentence about how you can prove to someone that each of the three translations of the A is a correct slide.
  Compare your sentences.

- Dot paper could be used to present examples and incorrect examples of *translations* of figures with different shapes, letters of the alphabet or numerals.
  Identify and name the examples that illustrate a translation or a slide and explain, orally or in writing, why that is.

- Dot paper is provided. The request is made to,
  Illustrate slides in two different directions with a trapezoid that has its vertices on the dots of the paper.
  Write instructions for someone that would illustrate the meaning of slide.

- **Rotation**
  An experience with different types of wheels means that most students are familiar with *turning* movements or *rotations*. *Rotations* are movements of a figure around a point. This *turning* point can be inside the figure, on a part of the figure or outside the figure. *Rotations* do not change the size or shape of a figure, but can change the appearance.

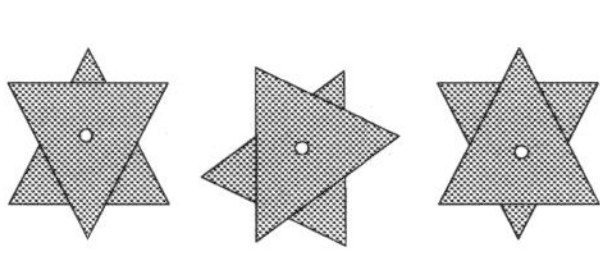

The following types of scenarios can be presented:

- *Pretend a square is attached to an analog clock, a bicycle wheel or a trundle wheel. The center of the square is attached to the fastener that holds the hands of the clock or to the bolt at the center of the bicycle wheel or the trundle wheel. One side of the square is parallel to the ground.*

  Prepare sketches of what you think the square will look like:
  a) After **¼ turn clockwise** *or when the hour hand points to the three.*
  b) After **½ turn clockwise** *or when the hour hand points to the six.*
  c) After **¾ turn clockwise** *or when the hour hand points to the nine.*
  d) After a **full turn** *of the hour hand or the wheel.*
  Write a sentence about your sketches and the different positions.
  Compare your sketches and your sentences.

- *Pretend one vertex of a square is attached to the center of a wheel or the center of a clock with the opposite vertex pointing up or to the twelve on the clock.*

  Prepare sketches of what you think the square looks like:
  a) *After* **¼ turn clockwise**.
  b) *After* **½ turn clockwise**.
  c) *After* **¾ turn clockwise**.
  d) *After a* **full turn**.
  Write a sentence about your sketches and the different positions.
  Compare your sketches and your sentences.

- *Pretend a square is attached to the top of the hour hand or to the top of a wheel with one side parallel to the ground.*

  Prepare sketches of what you think the square looks like:
  a) *After* **¼ turn clockwise**.
  b) *After* **½ turn clockwise**.
  c) *After* **¾ turn clockwise**.
  d) *After a* **full turn**.
  Write a sentence about your sketches and the different positions.
  Compare your sketches and your sentence.

These three different scenarios can be repeated with figures that have irregular shapes or with a simple diagram created by the students; like a sketch of a rabbit.

Creating new shapes.
The following requests can be made:
Place a square on top of a square. Use the center of the top square as a turning point and rotate it ¼ ***turn****.*
Trace around the outline of both squares.
Describe the shape of the new figure.

Place one isosceles triangle on top of another.
Use the center of the top triangle as a turning point and rotate it ¼ ***turn****.*
Trace around the outline of both triangles.
Describe the shape of the new figure.

- **Reflection**
  A *reflection* or *flip* is the movement of a figure about a line.
  The line of *reflection* can be on the side of a figure, on the *vertex* of a figure or aside from the figure.

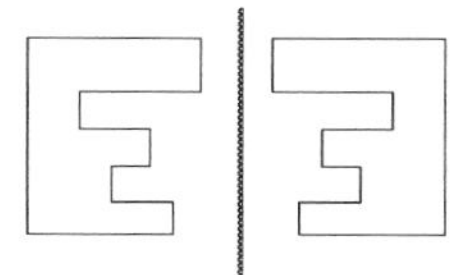

  Requests of the following type are presented:

  - *Draw a sketch of a three- or four-sided figure on dot paper with the vertices on the dots.*
    Draw a flip line on one side of the figure and then the flip image of the figure.
    How are the two figures the same? How are they different?
    Write a sentence.

  - *Draw a sketch of a three- or four-sided figure on dot paper with the vertices on the dots.*
    Draw a flip line on one vertex of the figure and then the flip image of the figure.
    How are the two figures the same? How are they different?
    Write a sentence.

  - *Draw a sketch of a three- or four-sided figure on dot paper with the vertices on the dots.*
    Draw a flip line beside the figure and then draw the flip image.
    How are the two figures the same? How are they different?
    Write a sentence.

  Reflection can be used to create unique shapes for possible patterns. For example, after printing a capital letter like an S, a *flip* line is drawn beside the letter and then the *flip* image. The new image created by the S and its *flip* image can then be used as part of creating patterns, or designs, or designs with patterns.

The geoboard is an excellent manipulative aid for experimenting with and for following instructions that deal with *slides* and testing predictions for *rotations* and *reflections*.[5]

The pegs or nails on a geoboard can be identified by ordered pairs of numerals. Directions to construct figures of specified shapes in a specific location on the geoboard can be given with sets of ordered pairs of numerals. Directions for performing a transformation on the figure that was constructed can then be added. For example, what directions are required to change a rectangle into a parallelogram or vice versa?

- **Symmetry**
  Examples from nature, art as well as pictures of familiar objects can be used to introduce ideas related to *symmetry*. There exist objects and pictures that make it possible to visualize and predict one half of a hidden part because these parts are *flip* images of the parts that are visible. These objects have a mirror line or a *line of symmetry*. A mirror placed along this line reveals the hidden part or the part behind the mirror.

  Folding can be used as a test for *symmetry*. For example, a square can be folded in four different ways and the hidden half can be visualized – a square has four mirror lines or four lines of *symmetry*.

  Lines of *symmetry* for figures can connect vertices, a vertex to a midpoint of a side, or the midpoints of two sides. A rectangle has two lines of *symmetry* and a parallelogram does not have any.

Types of Activities and Problems

- *Prepare a list of things around the classroom and the school that have lines of symmetry.*

  Suggest an appropriate title for the list.
  Why do you think the title is appropriate?
  The lists are compared and a composite list is displayed.

- *Collect pictures from magazines that show things that have at least one line of symmetry.*

  Make up an appropriate title for the collection.
  Why do you think the title is appropriate?
  The pictures are displayed along with a suitable title.

- *Draw several figures with more than one line of symmetry.*

  How many does each figure have?
  The pictures are displayed.

- *Take turns facing a partner.*

  Use your hands or hands and arms for actions that illustrate symmetry to your partner.
  Ask the partner to describe what is seen and state whether or not examples of symmetry are observed.

- *Do you think a face has a line of symmetry? Make a guess.*

  Carefully look at the two parts of your face and share your observations.
  Write a sentence to answer the question.
  Does everyone agree with your conclusion?

- *Which capital letters of the alphabet have a vertical line of symmetry? Prepare a list.*

  Which capital letters have a horizontal line of symmetry?
  Prepare a list. Compare your lists.

- *Try to think of a word or a name that has a vertical line of symmetry.*
  Try to think of a word or name that has a horizontal line of symmetry.
  Try to print two or three words or names that have a line of symmetry.
  Compare your words and names.
  Prepare a display for the library or the hallway with a title that tells viewers and readers that there is something special about the words they are looking at.
  Challenge them to guess what it is.

- *Use a felt pen and on a piece of paper print* **WATER BED** *in capital letters.*

  Take a test tube or a narrow glass filled with water and try to read what is printed by looking through the glass of water.
  Report what you saw and explain the reason for it.

  Make a guess about the printed words **BED BOX**. *Check your guess.*
  Was your guess correct? Why or why not? Write a sentence.
  Try to make a guess about **BOB BOXED OTTO**. *Write a sentence that explains the reason for your guess.*

Assessment Suggestions

Possible Types of Assessment Tasks

a) What would you say to someone who states that all *rectangular prisms* look the same?

b) Prepare a sketch of a *triangular prism*. Prepare a list of the number of *faces*, *edges* and *vertices* that are not visible or cannot be seen in the sketch. How do you know your answers are correct?

c) What would you say to someone who states that a *cube* is not a *rectangular prism*?

d) Name one object from outside the classroom that has the shape of a *rectangular prism* and one object that has the shape of a *triangular prism*.

e) Spread your second finger and your third finger as far as possible apart. Describe how you would *estimate* the *measure of the angle* between the two fingers.

f) What would you say to someone who suggests that it is possible for a triangle to have two *right angles*?

g) What would you say to someone who suggests that it is possible for a triangle to have two *obtuse angles*?

h) Which of the following are not true? Why is that the case?

- A *triangular prism* has two *faces* that are *parallel*.
- A *triangular prism* has four pairs of *parallel edges*.
- A *triangular prism* does not have any *right angles*.
- A *triangular prism* has three *rectangular faces*.

i) Try to sketch and name the figure:

- The figure has two pairs of *parallel sides*.
- The pairs of *sides* are not the same length.
- The figure has no *right angles*.

j) What would you say to someone who states that after a $\frac{1}{4}$ ***turn*** a square is not a square anymore?

k) Choose a capital letter of the alphabet and illustrate:

- A *slide*.
- A *flip*.
- A $\frac{3}{4}$ ***turn clockwise.***

Label each illustration.

l) Why are some people interested in *slides, flips* and *turns*? Write a sentence.

m) How would you explain to someone what a *line of symmetry* is? Write a sentence.

n) Make a sketch of something that has two *lines of symmetry*. How do you know the lines are *lines of symmetry*?

o) If you think a word from the list has a *line of symmetry*, show where you think it is and explain your thinking. **OX LET MOM BE**

p) Why are some people interested in *lines of symmetry*? Write a sentence.

Reporting

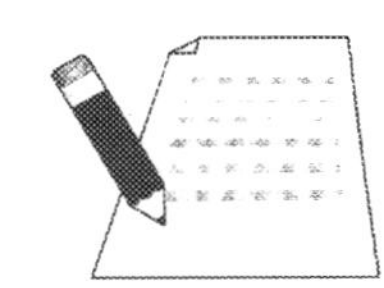

The responses to the assessment tasks make it possible to share information about students':

- Ability to *connect* ideas to settings outside the classroom.
- Understanding of some of the terminology and expressions that were used as part of the activities and problems.
- Ability to *visualize*.
- Ability and willingness to *communicate* and to explain what has been learned in one's own words.
- Ability to describe an *estimation strategy* for estimating the measure of an angle.

For Reflection

1) The curriculum includes the expression *spatial reasoning* without providing a definition or an example. How could or would you define *spatial reasoning* for someone? What example would you use to illustrate your definition?

2) How would you answer the questions, *'What is spatial sense and why is the development of spatial sense important?'*

3) Many adults will share the fact that they did not like anything that had to do with geometry when they studied mathematics. Some express an outright hatred for the topic. Why do you think that is the case?

4) During a presentation one parent shared the observation that in order to help her daughter with her mathematics, the parent had to relearn and memorize a long list of names for special solids. What would you say to this parent?

5) What would you include in a list of possible reasons for making geometry the first part or a very early part of the yearly mathematics program?

Chapter 8 – Patterns, Data Analysis, Probability and Game Setting

The general outcomes in the curriculum[1] for the topics that are listed in the title are related to fostering the ability to solve problems. The specific learning outcomes for these topics accommodate the key aspects of the *critical components* that students must encounter in a mathematics program:

communicating; *connecting*; *mathematical reasoning*; and *visualizing*.

Many of the teaching strategies that can be employed for the key ideas of these topics are suitable for *developing and applying new mathematical knowledge through problem solving*. Developing new ideas and procedures *through problem solving* can contribute to reaching the goals for students that are identified in the curriculum. These goals include:

confidence; *perseverance*, *risk taking*; and *exhibiting curiosity*.

Opportunities also exist to engage students in the use of their *imagination*.
ather than any specific content.

Patterns

Should patterns or the examination of patterns be presented as a separate unit or topic to students? It is very unlikely that anyone ever sets the study of patterns as a general or specific goal. The discovery, examination, description and use of patterns are part of the ongoing learning about mathematics. Mathematics learning could be thought of as the search for and study of patterns. The following examples selected from ***Making Mathematics Meaningful***[2] [3] illustrate the role of patterns as part of mathematics learning:

- Rational counting is based on increasing patterns. These patterns make it possible to examine any counting sequence and predict correctly:
  - The next number(s) and/or numeral(s).
  - The previous number(s) and/or numeral(s).
  - Missing numbers and/or numerals.

- Students make up their own rule for recording the answer when one factor is zero, test the rule and express it as an equation.

- From the entries of a chart with the headings of base, height and area for rectangular regions, students make up a rule for calculating area and express this rule as an equation.

- Knowledge of the pattern for the number of faces, edges and vertices for solids with flat faces enables students to make predictions.

The analysis of data includes observing and searching for possible patterns. Patterns that are observed are used to make predictions. Some of the predictions that are made may involve probabilistic thinking.

Data Analysis – Probabilistic Thinking

Analyses of organized data include the examination of bar graphs, double bar graphs and line graphs. The purpose of the dialogues that are included in this section is to illustrate the important role a teacher plays in accommodating the *critical components* that students must encounter and in reaching the *goals* that are listed for students. The examples also illustrate how learning about mathematics can contribute to language development, reading comprehension and the use of evaluative skills.

The conversations between teachers and students that are referred to or quoted in this book and many of the examples that are used are taken from lessons that were observed and video-taped.[4]

Data Analysis

Types of Activities and Problems

Parts of Representations – Missing Information

- **Importance of Titles:** Discussions about the examination of organized data can begin with looking at the importance and characteristics of titles. For example,
  - *How do you know what a graph is about?*
  - *Why is a title important?*
  - *What makes a good or an appropriate title for a graph, a story or a movie?*
  - *Make up or state an example of what you think is a good title for a story or movie? Why is that the case?*
  - *What do you think would make a title inappropriate?*
  - *Make up an example of a title for a movie or book that you think is inappropriate? Why is that the case?*

- **Missing Titles**: The speculation about missing information provides an opportunity for the use of imagination. All types of responses should be explained and justified.

  The titles of several graphs are hidden. After an examination of the graphs the request is made to make up appropriate titles for the graphs.
  - *Why do you think the titles are appropriate?*
  - *Is it possible to have more than one title that could be appropriate? Why or why not?*
  - *How do your titles compare to those created by the authors of the graphs?*

- **Missing Label**: A group of grade six students was shown a line plot entitled *Sunshine in Edmonton*.

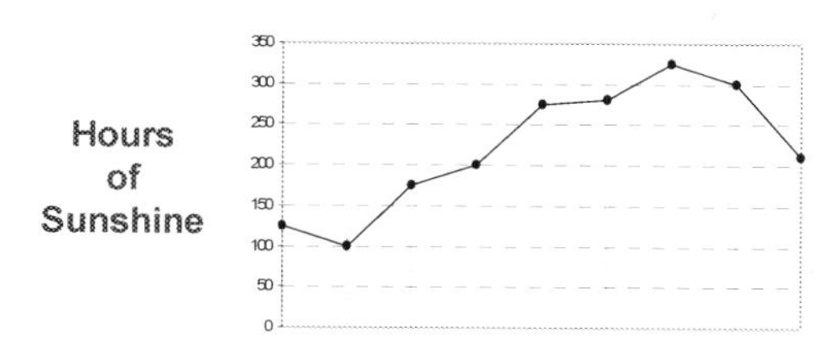

The vertical axis was labelled *Hours of Sunshine*. The nine entries for the hours were **125**, **100**, **175**, **200**, **275**, **280**, **325**, **300** and **210**, respectively. The label for the horizontal axis was missing.

After the students agreed that the first entry showed **125** hours and one group made the guess that the intervals along the horizontal axis could be weeks, the following exchange took place between the teacher (**T**) and the students (**S**):

T – *I would like you to estimate. About how many hours are there in a week?*
S/1 – *One hundred seventy five.*
T – *How did you figure that out?*
S/1 – *I multiplied seven times twenty five.*
T – *How would you calculate the exact number of hours in a week?*
S/2 – *Subtract seven.*
T – *What's the answer?*
S/3 – *One hundred sixty eight hours.*
T – *Do you think the intervals on the graph could be weeks?*
S/4 – *No.*
T – *Why not?*
S/5 – *Because half the time it is night.*
T – *If the intervals can't be days and can't be weeks, what do you think is shown?*
S/6 – *Months.*
T – *Discuss with your partner which months you think are shown. Be ready to provide a reason for your answer.*

The first group to report the correct response identified the first entry of the line plot to be for January. The dip was explained by suggesting that it is a result of February having fewer days. Number of hours of sunshine in the winter was explained by describing days as being clear and cold. The final request to the students consisted of making and justifying predictions for the next three months.

Representations with titles and labels for the horizontal axes but lacking any information for the vertical axes give students opportunities to speculate about possible numerical scales that they think could make sense for such displays of data.

- **Missing Labels**: A representation is shown that has a title but neither of the axes is labelled. As students work with a partner, they are to provide the missing information that they think makes sense to them and to explain, orally or in writing why that is the case. Providing the missing information may require research of some type.

Postal Rates

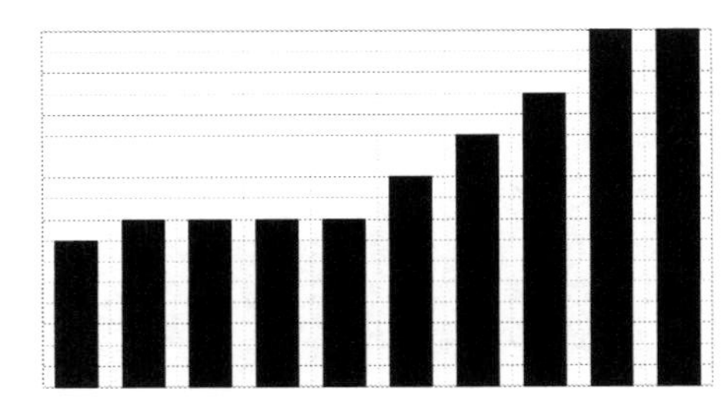

It did not take a group of students long to identify the values of a vertical scale for ten bars identified by the label *Postal Rates* for letters from quite a few years ago. The values for the first and last bar were **7** cents and **17** cents, respectively. It took longer for the students to find out the appropriate years for each bar and to suggest a calculation procedure for finding the average increase in rates for these ten years.

The students were also challenged to prepare responses and explain their thinking for the following questions,

- *How would you find out whether or not the average increase in postal rates is the same for different ten year periods?*
- *Do you think an increase in postal rates is likely or certain?*
- *Do you think it is possible to predict what the postal rate will be one or two years from now?*
- *Do you think postal rates will ever decrease?*

- **Missing Title and Label**: The provision of a data plot or a graph without a title and without any information for the horizontal axis provides an ideal opportunity to have students speculate and use their imagination. These speculations can result in some very interesting and lively discussions.

A group of students was provided with a double line data plot and the vertical scale showed **10.0**, **10.5** and **11.0**. Five intervals were marked off along the horizontal axis. The upper line started at **11.0** and decreased to about **10.5**. The lower plot started at about **10.3** and decreased to about **10.0**.

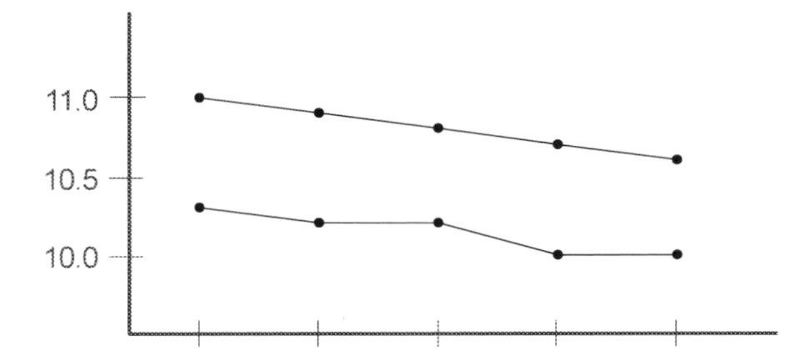

The request was made to speculate what this plot might be about and to write a few statements that are illustrated and supported by the data plots.

The decimals provided specific hints for the conclusions that were reached by all groups. One group assumed that the decimals referred to centimetres and millimetres and as a result interpreted the data plot to be about amount of rainfall for two different places. Another groups thought the decimals showed degrees and prepared a matching report. Two groups concluded that these could be times for swimming and running distances.

The students were told that the plots were generated from the times for the **100 metre** dash for males and females at five different Olympics. As part of a project, the following challenges were issued:

Try to find out which five Olympics were considered.
Describe the strategy used to find out.

Be ready to share your thinking about:
*Do you think the times for the **100 metre** dash will always keep on improving?*
Do you think the times for men and women will someday be the same?

- **Without Information**: A graph or a data plot is presented without a tile and without any information for the axes. The setting provides an opportunity for the use of imagination as information is created that makes sense for the parts of the display and the changes that are indicated.

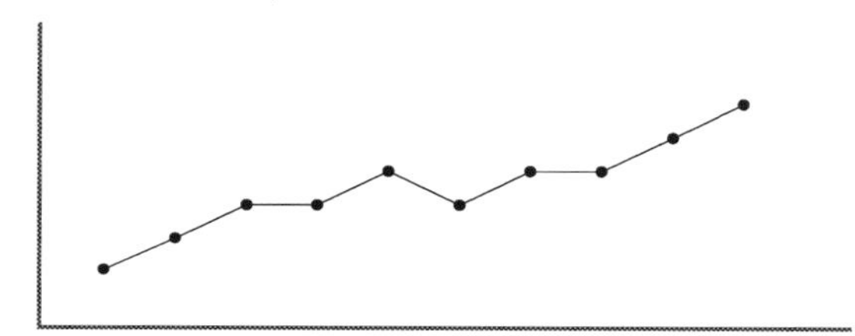

A data plot consisting of nine points indicating several increases, two instances of levelling off and one decrease was presented without any other information. Pairs of students were requested to consider at least one possibility for what the data could represent, to report reasons for this choice and then to try and explain the possible meanings of the changes indicated by the data plot as well as possible reasons for these changes.

Even if the titles suggested by the students in one classroom may have been influenced by recent current events, by newspaper clippings displayed on the bulletin board, or by discussions in other subjects, it is a satisfying experience to observe the use of imagination and how creative students can be when given the opportunity. Selected titles from different groups in this classroom included:

Greenhouse Effects
The Price of Homes
The Weight of a Baby
Tadpoles Hatching in a Pond.

As the groups gave their reports, they faced some very meaningful questions, challenges and further suggestions from their classmates. For example, there existed different opinions and possible reasons for the gains in weight for a baby, why the baby might lose weight and why the weight did not change.

The reporting was also turned into a guessing setting for one group. After the group revealed its title for the data plot, the classmates were invited to try to guess what the members of the group might have been thinking as they explained the changes indicated by the plot.

After the students were told that the data plot was created from the average speed of race cars at the Indy 500 over a ten year period, they were asked to consider responses for,

How would you go about finding out the years that are shown?
Do you think that over the years the average speed will keep on increasing?
Why do you think the average speed might have decreased?
Why do you think it remained the same on several occasions?

Collecting and Organizing Data

The references used by teachers include many suggestions for types of problems that involve collecting, organizing and analyzing data.

Types of Problems

While introducing teachers-to-be to problem solving settings that included collecting, organizing, interpreting and sharing of data, examples were selected from three categories: *Problems about Ourselves*; *Experiments and Investigations* and *Measurement Problems.*[5]

Problems about Ourselves could include questions or concerns about things students like to do and to watch, characteristics about themselves and generally things that are important to them. For example, what do they prefer to eat for different meals or, what are characteristics of a typical person their age? Pagni[6] in an article about human variability includes many ideas for tasks related to generic characteristics such as tongue, ear and thumb type; acquired characteristics such as preferences and extrasensory perception; and human characteristics that are measurable like reaction time and depth perception.

As part of a sequence of lessons about collecting, organizing and analyzing data one teacher began a session by asking students to generate questions that they would be interested in finding answers for.
The questions these students generated included:

- *If you were to join the armed forces which branch would you prefer to join?*
- *If you had a choice, which eye colour would you prefer?*
- *Are the pets you have in your home the kind of pets you prefer to have?*
- *If you had a choice, which would you prefer to own: Porsche, Mercedes, Lamborghini or Ferrari?*
- *Which do you prefer: slow dancing, fast dancing or no dancing?*

The origin of one question can be explained by the proximity of the school to a navy base. The data collected for the last question actually led to lessons in slow dancing in the gymnasium during lunch time.

Experiments and Investigations could include such questions as:

- *Is the number cube used for the game fair?*
- *How will a thumbtack land most of the time?*
- *What is the frequency of vowels on one page of the readers we use?*
- *Which consonants do we use most frequently?*
- *How many syllables do most of the words on a page of our reader have?*

Care must be taken when decisions are made about tasks that may involve gambling or things that can be associated with gambling. There are groups of people who object to the use of dice and playing cards.

It can be assumed that the data collection and the graphs for the task about the number of syllables would be very similar for different groups undertaking this task. In one classroom two interesting observations were made as groups shared their findings and a composite graph was constructed. Two groups made references to *rounding off* and *sort of averaging* or *sort of going to the middle* as they reported results to their classmates; an incident that lends itself to sharing more information about aspects of statistics.

The one result that was different was attributed to the fact that these students had examined a page of poetry. That raised the obvious question for a follow-up investigation,

Do all or most poets actually tend to use words with more syllables?

Measurement Problems could include such investigations as:

- *Say something about the measurements to the nearest centimetre of the wrists and the necks for several persons.*
- *If a rectangular region is traced around the fingertips of outstretched arms, head and feet, are most of the students you know shaped like a square or like a rectangle?*
- *Compare and say something about the pulse rate before and after exercise.*

Characteristics of Good Problems

One key characteristic of a good problem is that it is open-ended. The problem can be interpreted in different ways, several procedures for arriving at a solution may be possible, or more than one solution may exist. Such a problem provides maximum involvement on the part of the students and minimum teacher direction. It can lead to further problems and it may be possible to integrate the results into other areas of study.

The measurement problem about pulse rate before and after is one example of open-endedness. Students who have collected data and reported their results under such titles as *The Beat Goes On* and *You've Got the Beat!* have as part of the data collection considered such tasks as: walking up and down stairs; standing up and sitting down; chewing and not chewing gum; clenching and unclenching hands and writing numerals very quickly as opposed to very slowly.

Teachers are always on the lookout for opportunities to *connect* ideas and relate or transfer them to other subject areas or to settings outside the classroom. The reports by different groups in one classroom provided such an opportunity because the results caused the students to generate the following questions,

Why does our pulse rate go up and why does it not go up in the same way for everybody?
Is the pulse rate different for younger or for older students?
What is a normal pulse rate and how do you know whether or not it is normal for you?
What does it mean to be in shape as far as pulse rate is concerned?

These questions led to an invitation of a nurse to the classroom to solicit answers. There were some students who expressed interest in repeating the task after completing a mini-exercise program *to get in shape* or *to get in better shape.*

Collecting and Organizing Data

Conducting interviews, taking surveys and making tally marks are part of the data collection process. The task of organizing the data can involve decision making and problem solving if certain requirements or criteria have to be met:

- **Type of Data Plot**: The first decision is related to what type of data plot or graph might be appropriate or best suited for a report and display of the results.

- **Scale**: A decision about a numerical scale may need to be made.

- **Summary Statement**: The request for at least one summary statement of the most important result requires a focus on the main ideas of the problem that was solved.

- **Title**: The creation of a title that is appropriate and eye-catching can involve decision making as possible alternatives are considered.

- **Illustrations – Diagrams or Pictures**: The preparation of sketches or the selection of pictures from magazines that illustrate key ideas of the data display, the title, and/or the summary statement involves decision making.

Interpreting Data – Probabilistic Thinking

Initially a discussion is required about a data plot or a graph that leads to a conclusion that is synonymous with '*a picture is worth a thousand words*.' This discussion will point out the advantages of using data plots and graphs to illustrate and summarize results.

The requests to prepare a list of questions; a list of statements that are *true*, *false*, or *could be true* but the display does not tell us; or a list of statements that use probabilistic language which are to be part of the presentation to classmates requires thinking about the problem and the results.

A Data Plot or Graph is Worth a _____ Words

One possible setting can consist of displaying a data plot or graph and making the request to share or to record all of the information that can be abstracted by looking at the display. As the students make different statements, tally marks could be recorded. At the end of the discussion the number of statements that were made is announced.

As a plot or graph is examined it may be necessary to remind students to think about such terms as more, fewer, greatest, least, about the same, total and difference as they think about things to report.

Pairs of students could be asked to look at a graph or data plot about an issue that is reported on the news and they are asked to compose a report for a local newspaper or for a radio show. The report could be recorded and played for the classmates.

The conclusion for these types of tasks will be that indeed, representations like data plots or graphs 'are worth' many words and that is one reason why they are prepared and used. An examination of representations can result in many different observations and lead to conclusions.

Types of Questions

In one classroom each group of students solved a different problem but followed the same instructions for preparing the results for sharing. The request was made to generate two lists of questions. For one list, the answers to the questions could be found by looking at the graph that is part of the presentation. The questions of the second list had to deal with some aspect of the problem and the display, but the graph did not provide the answers.

Prior to sharing the problem, the organized data and the results with the group, another decision was required because the request was made to pose only three questions from each list; the three that the members of the group considered the best questions.

Types of Statements

The request is made to think about the problem that was solved and prepare three different types of statements that become part of a presentation to the group. Prior to making this part of an assignment all of the students could look at the same graph or data plot, try to author an example or two for each category and discuss these attempts as they are compared.

- **Statements that are <u>True</u>**: The listeners can identify and label these statements as true by looking at the data plot, the title or the summary statement.

- **Statements that <u>Could be True</u>, but the display does not show it**: This list can also include statements that are true, or the students know them to be true, but the data plot does not confirm it.

- **Statements that are <u>False</u>**: Based on what the data display shows, the students will recognize or conclude that the statements are false.

Probabilistic Language

As a graph or data plot is examined, the request is made to make up sentences about the data that is part of the display and use the words:

likely *unlikely* *impossible* *certain.*

The category *impossible* may require some guidance. For example, is it *impossible* to tell from the data plot that a given statement is *likely*, *unlikely* or *certain* or is a result or action included in a statement that is not possible?

After the statements have been discussed and compared, the words *probable* and *improbable* are introduced for *likely* and *unlikely*, respectively. The new words are included in future presentations about collected data to classmates.

The introduction of this type of language along with the use of ratios will lead to understanding the ideas and skills related to probability. Probability is how likely or how probable it is for something to occur:

How likely is it or, what is the probability of a thumb tack landing on its side?

Assessment Suggestions

The categories for assessment tasks can include questions, comments and reactions about:

- Parts of a data plot or graph.
- Interpretation questions about data that are displayed.
- Ability to elicit suggestions for how students might collect and organize data to solve a problem that is presented.
- Ability to interpret parts of a data plot or graph.
- Ability to identify and extend patterns; extrapolate.

Possible Types of Assessment Tasks

Accommodating Responses – Assessment Items for Patterns:

Since repeating patterns can easily be changed to increasing patterns and vice versa, there are different answers possible for the question about patterns that asks, *what comes next*? Some authors of assessment items seem to make the assumption that there is just one way, or just one logical way to extend a pattern. On tests the request is made to '*extend a pattern in a logical way*' and the answer key identifies one response as being correct.
For example:

Continue in a logical way: M T W T _ _ _

Students who have experienced open-ended settings, learned new ideas *through* problem solving and are confident risk takers will be able to extend this sequence in many different logical ways; other than thinking of the days of the week, which was identified as the correct answer for this example. If the intent is to have students identify one specific response, in this case the days of the week, very specific instructions are required.

Parts of a Data Plot or Graph
A request is made to react, orally or in writing, to scenarios of the following types:

a) A group of students drew a bar graph that showed which of the four seasons the people in the classroom preferred. The title the students made up for the graph was:
This Bar Graph Shows Which Season the Students in this Classroom Like the Best.

b) The students who drew the graph about which of the four season they preferred labelled the horizontal axis:
Season 1 Season 2 Season 3 Season 4
The Four Seasons

c) The concluding statement on the graph about which of the four seasons the group preferred was: *We like the four Seasons.*

Interpretation Questions
Common types of interpretation requests involve comparisons (*most, least, about the same, the same*) and arithmetic skills (*find the total, find the difference).*
Other types of requests can involve the classification of statements and the evaluations of questions.

A request is made to react, orally or in writing, to scenarios of the following types:

a) Describe each of the following statements as *true*, *false* or, *it could be true,* but the data plot or the graph does not show the required information.

continued next page

Consider the statements for the graph:

The Favourite Season for the People in this Classroom

There are boys and girls in the classroom.
The graph shows only the results for the boys.
The teacher is part of the graph.
The principal is part of the graph.
More girls than boys prefer spring.
Nobody was absent when the data were collected.

b) Rather than using the categories *true*, *false* and *could be true – but the information does not show or tell us*, requests are made to examine statements and classify these as one of the following:

Likely or *probable*. *Unlikely* or *not probable*. *Certain*. *Impossible*.

Consider the statements for the graph:

The Favourite Season for the People in this Classroom

A boy named Dylan likes the Spring season the best.
The teacher is part of the display.
One student was absent when the data were collected.
The Spring and Summer seasons are preferred by most people in this classroom.
Some students liked the Summer season the best.
Nobody in the classroom liked the Fall season.

Explaining a Problem Solving Procedure

The school needs help in trying to figure out what outdoor sports equipment to order for the intermediate grades.

How could you go about finding the information that is needed and how would you prepare your answer to make it easy for the people in the office to interpret the results.

Explain your strategy of identifying the information that is required, how the information will be collected, and how it will be organized.
Make a sketch of what your presentation might look like.

Interpreting a Data Plot or a Graph

A data plot or part of a line graph that shows two or three rises, two dips and two level parts without a title and without any labelling is prepared. Two possible assessment scenarios include:

a) **Data Plot without Information**: An invitation is issued to use imagination and to suggest what the display could be about. If a suggestion is made, it is followed with a request for a title; possible labels for the axes; and a brief paragraph about what the display shows or illustrates.

b) **Data Plot with a Title**: After a title is suggested for a display, the request is made to keep it in mind and to write a paragraph explaining what the parts of the representation tell about the title. Suggestions for titles could include:

A Walk with My Dog *Riding My Bike* *Sidewalk Skateboarding*
Playing My Guitar *Going to My Grandparents' House*

Does the paragraph that is created indicate an awareness of:

- a major general conclusion that is illustrated?
- each of the changes that are part of a representation?
- how the changes relate to one another?

Reporting

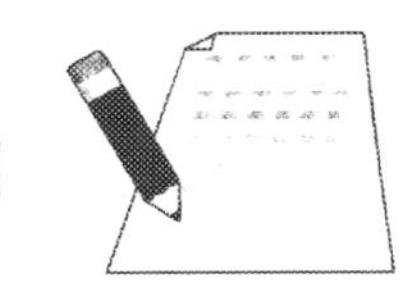

The type of assessment tasks make it possible to share comments about:

- Understanding of the role and importance of an appropriate title.
- Knowledge of the purpose of labelling the axes.
- Ability to interpret questions and statements about data organized in a data plot or a graph.
- Ability to use data plots or graphs to find the answers for problems.
- Indicators of *confidence, willingness to take risks* and *use of imagination.*

For Reflection

1) What is one example that you could use to illustrate that teaching about mathematics can contribute to language development?

2) What is one example that you could use to illustrate that teaching about mathematics can contribute to reading comprehension?

3) What is one example that you could use to illustrate that teaching about mathematics can contribute to the development of evaluative skills?

4) What example or examples could you use to explain and illustrate to someone that the development of conceptual understanding of mathematics is dependent upon the orchestration of discussions and the accommodation of students' responses?

5) What sample responses and explanations would you use to illustrate to someone that it is inappropriate to have one correct answer in mind for the following request:
Continue in a logical way: **M T W T** _ _ _

Game Settings and Games

Games as Part of Teaching about Mathematics

The main reason for the inclusion of several games in the book is to show examples that:

- meet as many aspects of the *critical components* and *goals* of the curriculum as possible.
- meet criteria for good games for the mathematics classroom.
- illustrate the importance of the role of a teacher in orchestrating discussions during a game or after a game is played.

Mathematical Processes

Many of the *critical components* identified in the mathematics curriculum can be accommodated in game settings. *Communication skills* can be fostered. Mathematical skills and ideas are *connected* to new skills and ideas. Opportunities for *mental mathematics* and *estimation* can be provided. It is possible to pose questions or to assign tasks which create learning *through problem solving*. Aspects of *mathematical thinking* that can include: *sense making*; *getting* oneself *unstuck*; *identifying errors* in thinking, moves or use of materials; and opportunities to *try different strategies* can all be accommodated. Some game settings can contribute to fostering the development of *visualization*.

Selected Criteria for Appropriate or Effective Games

To be consistent with the students' experiences games should have rules and there should be a winner or winners. However, it is possible to create settings where the emphasis is not on winning and all participants get rewarded in some way.

Finding games that meet all of the criteria for good games that are listed may be a difficult task. The assumption is made that the more of the suggested criteria are met, the more effective or the better the game.[2][3]

Criteria for effective games can include:

- The rules should be simple. Time required to give lengthy explanations can lead to boredom and confusion.

- There should be quick action. Quick action helps to maintain interest and allows for coverage of many ideas or skills or for repetitive coverage of these ideas and skills. Slow action or lengthy waits between moves or opportunities to participate can result in boredom and loss of interest.

- Few and simple pieces should be required. Many pieces require lengthy set-ups and this can easily distract from the intended goals of a game. Any loss of pieces or unintentional rearrangement of pieces during a game can result in breaks in the action and a loss of interest in a game.

- Games with a chance outcome give everyone an equal opportunity to win. A focus of fighting one another or winning a war should be avoided. With all the violence that exists, references to war seem inappropriate. Being a winner or a winning team should or could simply mean having been lucky. An inability to predict the winner can be reinforced by using *Lucky* as part of the name for a game. Finishers could be called first winner, second winner, etc. and every player or every group should be rewarded in some way. For example, a **10**, **9**, **8**, **7**, etc. point system can be adopted to reward all players.

  A chance outcome for a game does not imply that possibilities to think, to solve problems and to ponder strategies while moves are considered and made during a game do not exist. The strategies that are employed will not determine the final outcome of a game. However, there exist appropriate types of games that allow those who do solve problems and develop strategies to become better at playing these games.[7]

- The game should be a learning experience. An incorrect response should not be punished in any way. Opportunity should exist to correct and to keep participating.

- The game should allow for questions and discussions and responses to certain types of questions during the game. Discussions could focus on invited comments to questions or incidents of the following type:
  - *If you did 'this' or made 'this' move, why is that the case?*
  - *What do you hope will happen next? Why?*
  - *I heard someone sigh. Why do you think that happened?*
  - *Some students said, 'Yes!'- Why do you think they did?*
  - *Show on your fingers what number you would like to have come up next? What do you think we can say about students who want that number to come up next?*

- Good games are flexible. They can be used in different settings – large groups, small groups, with a partner or even in a solitary setting.

- Good games can easily be modified to accommodate new topics and new learning outcomes.

- Physical actions are desirable in some settings, but in the mathematics classroom they can and they do distract from the purpose or the outcomes of a game. Throwing bean bags, tossing rings or casting fishing rods will make students focus on the action rather than the problems that were to be solved.

This list of criteria is suggestive of the existence of games that are not appropriate or are educationally unsound for use in the mathematics classroom. Such games do exist. These types of games might:

- have too many parts or pieces to keep track of.
- have rules that can be quite complicated and require lengthy explanations as well as examples for strategies to be used during the game.
- punish inappropriate or incorrect responses by students by taking them out of the game.
- end with few or just two participants.
- not let players know during a game whether or not their responses are correct.
- make it possible to predict who in a classroom has the best chance of winning because of the special skills that someone has.
- be considered inappropriate by some groups of people since they involve or make reference to aspects of gambling.

Sample Games

It is not suggested that the games that are described are necessarily the best games that are available, but aside from the fact that important learning outcomes and goals are accommodated, observations collected over the years indicate that these games were well received by students. Indicators of this reception are not just the reactions that were observed while students played these games, but also included the requests to play the games again.

The names for these games are arbitrary. Students can be given the opportunity to discuss and decide on names that they would like to adopt for games.

Decimal Number Sense: Greatest Decimal

Setting: All students.

Materials: An envelope with ten disks or pieces of cardboard, each piece showing one of **0** to **9**.
Each player has a game sheet with several of the following markings for each round of the game: Two boxes labelled **T** for tenths and **H** for hundredths, one box for a discard.

Round 1

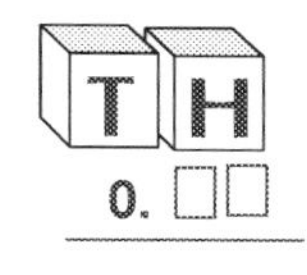

The goal of the game is to try and be lucky enough to end up with a name for a decimal greater than those recorded by the other participants.

Procedure:

For **Round 1**, a numeral is drawn from the envelope and is announced. Each player records the numeral in one of the three boxes. The numeral is returned to the envelope. This procedure is repeated two more times with one opportunity to discard a numeral judged not to be in the best interest as far as the goal of the game is concerned.

Possible Scoring Systems

- **Addition and Ordering of Decimals**: After five rounds the students add the decimals they recorded for each turn. The top sums, five or more, are recorded on the chalkboard. Points could be awarded. For example, five points for the greatest sum, four for the next greatest, and so on.
- **Subtraction and Ordering of Decimals**: After each round, the students calculate the difference between the greatest decimal reported and the decimal they recorded. After five rounds, the differences are added. Points are awarded. For example, five points for the least difference, four for the next least, and so on.
- **Ordering Decimals**: If the students have not been introduced to the addition and subtraction of decimals, the students are asked to report their decimals. These decimals are recorded on the chalkboard in order from greatest to least. Points are awarded. For example, five points for the greatest, four for the next greatest, and so on. Students record their points for **Round 1** and the game continues with **Round 2**. After five rounds, players calculate the total for the number of points they were lucky enough to obtain.

Possible Variations

- The purpose of the game is changed to **Least Decimal**.
- The numerals drawn are not returned to the envelope.
- Another place value position, **TH** or Thousandths, is added.
- Another discard box is added.
- The goals of the game could be changed to **Greatest** or **Least Measurement** or **Greatest** or **Least Amount of Money**.
- The game can be adapted by adding boxes for the ones or the ones and tens place.

Basic Facts: Multiplication or Division

Setting: Two players.

Materials: Decks or parts of decks of basic multiplication or basic division facts
A game board consisting of a linear arrangement of thirteen square shaped regions.
The squares on either end are labelled **Goal** and the square in the middle is labelled **Start**.[8]

The goal of the game is to advance a marker (puck; soccer ball) into or through an opponent's goal.

Procedure:

A goal is assigned to each player. A coin could be tossed or a guess could be made about a recorded numeral to decide who makes the first move. A marker – puck, ball – is placed on **Start**. The deck of flashcards is placed next to the game board.

The starter draws the top card from the deck, announces the answer and advances the marker toward the opponent's goal. For a game with basic multiplication facts the marker is advanced according to the number of ones in the product. For the basic division facts the marker is advanced according to the quotient.

Turns are taken until the marker ends up in a goal or goes through it. Whenever that happens, the player who scored receives a goal or point. The marker is returned to the Start position and the procedure is as it was at the beginning of the game.

The duration of a game can be determined by setting a time limit or by stating how many times the players are to go through a selected set of reshuffled flashcards. The selection of the cards could be based on a predetermined need for exposure and practice.

The players are requested to record on a sheet any basic fact equations they had incorrect or that required two tries to get an answer. These sheets may reveal error patterns that can become the focus of an intervention **IEP**.

Number Sense: Decimal Target

Setting: All students.

Materials: An envelope with ten disks or pieces of cardboard, each piece showing one of **0** to **9**.
Each player has a game sheet with several of the following markings for each round of the game:
Two boxes labelled **T** for tenths and **H** for hundredths, one box for a discard.

Round 1

T H

Target 0.☐☐ 0.☐☐

☐ **How far off the target?** 0.☐☐

Two students could be given the opportunity to create the target by entering their favourite single digit numeral into one of the boxes. Their names are printed below the target.

The goal of the game is to try and get as close to the target as possible.

Procedure:

After a target has been identified, for example **0.75**, and before the game starts, the following types of requests could be made,

- *Name a decimal that is very close to **0.75**.*
- *Name a different decimal name that is just as close as the last answer.*
- *Name a decimal that is less than **0.75**, and is far away from it.*
- *Name a decimal greater than **0.75** but less than **1.00**.*
- *Name several pairs of decimals that are the same distance from the target.*

The playing procedure is the same as for ***Greatest Decimal***.
After each **Round** the students are asked to calculate the difference between the **Target** and the decimal that was recorded.

Possible Scoring Systems

- **Ordering Responses**: All results of the calculations are recorded in order from least to greatest on the chalkboard. Points are awarded, i.e., five for being closest to the target, four for the next closest, and so on. After several rounds the points are added.
- **Calculating Totals for the Differences**: After several rounds, the differences are added for each round. Points are awarded, i.e., five points for the least difference, four for the next least, etc.
- **Calculating the Average for the Differences**: After several rounds, the differences are added and the average distance from the **Target** for those rounds is calculated. Five points are awarded for the lowest average, four for the next lowest, etc.

Possible Variations

- The goal of the game is changed to trying to stay as far away as possible from the target.
- The target is thought of as a distance in centimetres or as an amount of money.
- The target and the game sheet are extended to include thousandths for each round.
- The target and the game sheet are extended to include boxes for the ones or the ones and tens place.

Sums and Estimation: Target 3.33

Setting: All students.

Materials: Two envelopes, each with nine disks or pieces of cardboard - each piece showing one of **1** to **9**.
A game sheet for each with the following markings: **0. T H**

1) 0. _ _
2) 0. _ _
3) 0. _ _
4) 0. _ _
5) 0. _ _

Total ________ **How far off?** ______

The goal of the game is to try and get as close as possible to **3.33.**

Procedure:

One disk or one piece of cardboard is drawn from each of the two envelopes for each of the five entries indicated on the game sheet. The numerals are announced. After each announcement, the students have to decide which of the two numerals to enter in the tenth place and which to enter in the hundredths place. After the five draws the total and then the difference between the total and **3.33** is calculated.

Possible Scoring Systems

One of the procedures suggested for the game ***Decimal Target*** can be adopted.

After a round, students can be given the opportunity to describe and compare the different strategies they used during the game.

- *Explain some of your thinking during the game?*
- *What did you try to keep track of the decimals and how did you do it?*
- *How would you change your strategy if the goal of the game is changed to trying to keep as far away as possible from the target **3.33**?*

Possible Variations

- The goal is changed to trying to stay as far away as possible from **3.33**.
- Thousandths are added and the target is changed to **3.333** and three digits are drawn for each entry.
- Assuming that the target is a measurement in metres provides opportunities to think and talk about decimetres, centimetres and millimetres.

Strategies: Right or Right On!

Setting: All students, teams or partners.

The goal of the game is to try to identify a recorded two-digit mystery numeral made up of different digits in as few guesses as possible and to elicit the response, '*Twice **Right On**!*'

Procedure:

The request is made to make guesses about the numeral that has been written on a piece of paper. Each guess is recorded on the chalkboard under **Tens** and **Ones**. Entries under **Right** and/or **Right On** share information about the accuracy of the guesses:

- **Right** tells that a digit has been identified correctly, but it is in the wrong place value position.
- **Right On** tells that a digit has been correctly identified and it is in the correct place value position.

To keep a record of the guesses and responses the following chart is shown on the chalkboard:

Numeral Guessed		Responses	
Tens	Ones	Right	Right On

The numerals **0** to **9** are listed on the chalkboard. Any numerals that are eliminated as a result of the responses to guesses are crossed out.

Prior to start of the game the students could be invited to record a guess about how many tries they think it will take to identify the recorded numeral. These guesses are revisited and reactions are shared at the conclusion of the game.

After each guess and each response during the game, the students are invited to state all of the things they think they know about the numeral to that point. One student is asked to suggest the next guess and to explain the reason or reasons for making this guess. The discussion that takes place during this game illustrates the value of students explaining and defending their thinking. Students will correct one another and they will learn to realize that different approaches are possible.

At times it may be tempting to guide the students' thinking and to share with them a strategy or strategies thought to be more worthwhile than those suggested. Some of these strategies may surface during a discussion at the end of a game when an opportunity is given to reflect or it may take several games before strategies are refined. It should be kept in mind that at times a lucky guess may lead to the correct numeral more quickly than a strategy that is deemed appropriate or most appropriate.

Differences in opinions with regards to strategies are usually noticed after a first announcement of **Right** or **Right On**. In response to '*one Right*', some students will suggest reversing the numerals that were guessed, some will want to use one of the numerals that yielded this response as part of the next guess, while others will suggest the use of two different numerals. The discussion can be quite lively and may require careful non-judgmental orchestration. .

After the game has been played several times, teams or pairs of students could face each other. A game sheet for such a setting could have the following headings:

My/Our Number is: ___________ **Possible Digits: 0 1 2 3 4 5 6 7 8 9**

Opponent's Guesses		Responses		My Guesses		Responses	
Digits				Digits			
Tens	Ones	Right	Right On	Tens	Ones	Right	Right On

Possible Variations

- A three-digit numeral is recorded.
- Digits may be repeated.

Estimation: Lengths of Segments

Setting: All students.

Materials: Several segments that are joined at the endpoints and point in different directions are drawn on a piece of cardboard.

The goal of the game is to employ different strategies to estimate the lengths of different paths in centimetres and to gain familiarity with the centimetre as a unit of measurement of length.

Procedure:

The students are assigned to one of five groups. For each **Round** of the game different segments that are joined are shown to each group. For example,

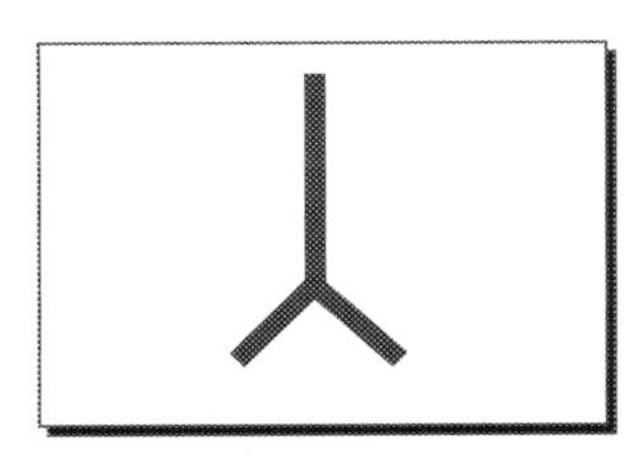

Round 1 **Round 2**

The students are asked to estimate the total length of the segments in centimetres. Every student is requested to record an estimate.

Before one estimate from each group is recorded, the members of the group have to reach consensus about the estimate they think best represents the estimates made by the members of the group. The students could decide on the estimate that is *sort of in the middle* of the estimates, or they could calculate the average of the estimates and round this to the nearest whole centimetre.

Charts on the chalkboard can be used for record keeping.
One chart can show the estimates each group has decided on for each round. Another chart can show the points earned by each group.[9]

Estimates

	Groups					
Round	**1**	**2**	**3**	**4**	**5**	
1. About	__	__	__	__	__	**cm**
2. About	__	__	__	__	__	**cm**

Points

	Groups					
Round	**1**	**2**	**3**	**4**	**5**	
1.	__	__	__	__	__	**cm**
2.	__	__	__	__	__	**cm**

The total length of the segments is measured to the nearest centimetre and the differences between the estimates and the measured length can be used as the score for each group. After several rounds the groups could be asked to calculate the average difference between the actual lengths and the estimates.

Rather than using the differences between the actual length and the estimates as scores, a **10 – 9 – 8 – 7 – 6** point system could be used where the highest score is assigned to the estimate that is closest to the actual measurement.

After each round a few players are given the opportunity to share their *estimation strategies.*

Guessing, Data Analysis and Strategy: How Many are Hidden?

Setting: Three players.

Materials: Three counters for each player.

The goal of the game is to guess how many counters are hidden in three hands.

Procedure:

The players hold the three counters and both hands behind their backs. Three, two, one or zero counters are transferred into one hand which is then closed, brought to the front and placed onto the table. The other hand is kept behind the back.

One player gets first turn with trying to guess how many counters there are in the three hands on the table. Then the other players take turns with the restriction that the number guessed by any player may not be used again.

The players open their hands to see whether or not a correct guess was made and a winner can be declared. If a correct guess was not made, it is called a draw or tie.

Turns are taken to be first, second and third to make a guess as the procedure of selecting counters and bringing one hand forward is repeated each time.

The setting is well suited for preparing several frequency tallies or bar graphs to keep a record of the outcomes:

- Which numbers were guessed? **0 1 2 3 4 5 6 7 8 9**
- What were the winning numbers? **0 1 2 3 4 5 6 7 8 9**
- What were the outcomes of the games? **Win Draw**
- Which guess resulted in the winner? **First Guess Second Guess Third Guess**
- Who won? **Names of Players: A B C**

Analysis of Collected Data - As the frequency tallies or bar graphs are examined the request is made to,

Suggest how the data and the results might determine or change your playing strategy for future games.

A cumulative frequency tally and a bar graph are constructed for,

Which Guess Resulted in the Winner?

The following types of questions can be posed,

- *What do these results indicate or mean to you? Why is that the case?*
- *How might it be possible for the players who guess first and second to change these results?*

If it is suggested that the players who guess first and second could make guesses that according to the number of counters they hold could not be possible, the students should be asked,

- *Would that be fair? Why or why not?*
- *Should it be allowed? Why or why not?*

Possible Variations

- Each player uses four counters.
  What is the same about this game and how is it different?
- Four players play the game with three counters and then with four counters.
  What is the same and what is different about the settings?

Use of Imagination: Inventing Games and Game Settings

The request to invent or make up a game provides the opportunity to use imagination and to be creative.[10] This request can include the following decision making aspects:

- **A title**: What do you think is a good title, one that might make other students curious about your game or tells them a little about the game?
- **A general goal**: What should those who will play the game keep in mind? What is the main goal or purpose of the game?
- **Specific goals**: Does the game have any specific goals? Are there some things that players should keep in mind while playing the game? How is a winner determined?
- **Rules**: What are the rules that need to be followed?
- **Decorations**: Are there some ways to decorate your game that would show players what the game is about? Are there some ways to illustrate the title or some of the action that is part of the game?

When a game is shared with classmates and the rules are explained or tested, the questions that are posed will likely result in revisions of some rules. Additional rules may be added. Rewriting and revising the rules provides opportunities for further reflection.

Accommodating Responses:

The observations and the data collected during these types of tasks indicate that students can be very creative, indeed. For many students some guidance and a few reminders will be required to have the game deal with as many aspects of mathematics as possible, i.e., mathematical operations and language. The students should be reminded or challenged to invent a game setting that does not make a reference to war or violence. Some students may prefer to attempt the task of creating a game in a cooperative setting.

Possible Materials

- **Activity sheets**: Any activity or practice sheet for any topic can be used and this can include tasks for basic facts, algorithmic procedures, measurement ideas, spatial sense.
- **Familiar game boards**: The students are asked to think of a game board they have used, have seen someone use, or would like to use. If they were to design a different game or make up new rules for the game board, what would they be?
- **Unfamiliar games and boards**: The students are asked to look at the board and pieces of a game, i.e., Chess or Backgammon, and invent their own game for the setting. What would your rules be for your game on this game board?
- **Pieces of cardboard and pictures or drawings:** To create a deck of cards the students use pictures of animals, objects, 3-D figures or 2-D figures. Rules are made up for at least one game for the deck of cards.

Many opportunities exist to foster the development of *communication* skills as the game that has been created is explained to the group or to players who are trying it for the first time. The questions that are asked during the orchestration of these settings can result in having students *think* as well as *think about their thinking*. Questions that are posed and reactions from players can result in revisions that can contribute to *advancing thinking*.

Assessment Suggestions

Since there are no specific goals related to playing and inventing games, the assessment information that can be collected and shared is of a general nature. The indicators that can become apparent during the activities related to games can provide valuable information about several desirable characteristics that are related to success with mathematics learning.

These characteristics can include possible indicators of:

- *Willingness to talk.*
- *Willingness to try strategies as well as different strategies.*
- *Thinking* and *thinking about thinking.*
- *Flexible thinking.*
- *Curiosity.*
- *Spontaneity.*
- *Ability to connect.*
- *Use of imagination.*
- *High self-esteem* or *confidence.*

Reporting

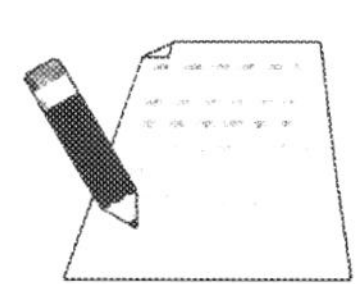

The observations that are made as: games are played, strategies are discussed and games are invented can provide information about the *critical components* that students must encounter in a mathematics program.

Data can become available about:

- *Communication skills*: ability to talk and write as strategies are explained or created.
- *Mathematical reasoning*: drawing conclusions about strategies or data collected about strategies; making up meaningful rules.
- *Visualizing* and use of *imagination*: designing rules and moves for a game or for a game on a game board.
- *Confidence*, *risk taking* and *perseverance*: sharing invented rules and revising the rules.

For Reflection

1) What points would you include in a presentation about the role of games as part of learning about mathematics?

2) What would you say to someone who makes the point that games should not be played in the mathematics classroom because these games can be played at home?

3) How would you respond to the claim that since Chess is a game that is mathematical the game should be part of mathematics teaching and learning?

4) What is your response to a parent who asks, Are there any special or very effective games that can be played in the home which would help to support the mathematics learning in the classroom?

5) What example or examples would you use to illustrate that one of the key roles for a teacher is the skilful orchestration of discussions?

6) An Issues and Ideas column entitled, *It's time to take a new approach to math*[11] includes the result of *a 2007 report from the Canadian Council on Learning that about 30 per cent of Canadian parents had hired tutors to assist their kids with math.* To address this concern, the author lists several questions about mathematics teaching and learning. What are your reactions to the following entry on that list: *Should we be completely reinventing the classroom environment when it comes to math?*